One Evening
Electronics Projects

HOWARD W. SAMS & COMPANY
HAYDEN BOOKS

Related Titles

**For the retailer nearest you, or to order directly from the publisher,
call 800-428-SAMS. In Indiana, Alaska, and Hawaii call 317-298-5699.**

One Evening Electronics Projects

Second Edition

Calvin R. Graf and Richard S. Goss

HOWARD W. SAMS & COMPANY

A Division of Macmillan, Inc.
4300 West 62nd Street
Indianapolis, Indiana 46268 USA

International Standard Book Number: 0-672-22551-4
Library of Congress Catalog Card Number: 89-63718

Acquisitions Editor: *Greg Michael*
Development Editor: *Jennifer Ackley*
Illustrator: *Don Clemons*
Cover Illustrator: *Ric Harbin*
Composition: *Cromer Graphics*

Printed in the United States of America

Trademark Acknowledgments

Contents

Preface .... *xv*

Acknowledgments .... *xvii*

1 **Construction Assembly Hints** *1*

Background of Projects *1*
    The Probevolt *2*
    LED Voltage and Polarity Indicator *2*
    A Solid-State Telephone Bell *2*
    The Poweralert *2*
    A Light-Sensitive Audio Oscillator *2*
    Polarity-Sensing Continuity Tester Using LEDs *3*
    A Light-Intensity-Level Wheatstone Bridge *3*
    An Audio Continuity and Voltage Tester *3*
    The LED in Motion *3*
    Talking Over a Light Beam *3*
    Audio Continuity Tester for Wide Resistance Range ... *4*
    The Lightstalker *4*
    A Visual Telephone Ringer *4*
    The Neon Telelite *4*
    Add Music on Hold to Your Telephone *5*
    Design Your Own Transistor Audio Amplifier *5*
Circuit Diagram Electronics Symbols *5*
    Component Values *6*
    Resistor Color Coding *6*
Electronic Component Mounting *8*
    Breadboard *9*
    Perforated Boards *9*
    Printed-Circuit (PC) Board *10*
    Solderless IC Printed-Circuit Board *11*

Cardboard *11*
Wiring and Soldering Methods *11*
Point-to-Point Wiring *12*
Test Clip Wiring. *13*
The Soldering Iron *13*
Soldering Wires and Components *14*
Soldering IC Chips and LEDs *15*
Unsoldering Components *15*
The Desoldering Tool *15*
The Solder Wick Absorber *15*
Packaging the Project *16*
Metal Cases *16*
Plastic Cases *16*
Felt-Tip Marker *17*
Bandage Boxes *17*
Lettering Your Project. *17*

2   The Probevolt Voltage Detector *19*

Circuit Analysis *19*
The Light-Emitting Diode. *21*
Construction. *26*
The Case *26*
Parts List *26*
Tip of Probevolt. *27*
Assembly *27*
Using the Probevolt *28*
Voltage Sensing *29*
Sense Battery Condition. *29*
Polarity Sensing. *31*
Response to Different Waveforms. *33*
The Moving LED *33*
Circuit Loading *34*

3   LED Voltage and Polarity Indicator *35*

Operation. *35*
Circuit Analysis *35*
Wide Voltage Range Operation *37*
Construction. *38*

The Parts *38*
Packaging *39*
Labeling. *39*
Using the LED Indicator *40*
Check Wiring *40*
Testing the Unit. *40*
Voltage Sensing *41*
Polarity Sensing *41*
Operation on AC, DC, and Audio Voltages *42*
AC Voltages *42*
The Moving LED. *42*
DC Voltages *43*
Audio Voltages *43*

4   A Solid-State Telephone Bell: The Sonabell   *45*

Operation. *45*
Circuit Analysis *46*
Piezoelectric Sounder *47*
Optoelectronic Connection. *48*
The Lady Bug *48*
Make Your Own Coupler *50*
A Ringing LED. *51*
Construction. *51*
Connection to Telephone Line *52*
Using the Sonabell. *52*
Additional Uses for the Sonabell *53*

5   The Poweralert: A Line Voltage Monitor   55

Operation During Power Failure *55*
Circuit Analysis *56*
Construction. *57*
Testing Operation *58*
Wake up During Power Failure. *58*
Additional Uses for Poweralert. *59*
Locate Circuit Breakers *59*
Freezer Alarm. *59*
Intruder Alarm *60*

6 A Light-Sensitive Audio Oscillator: The Sonalight

6 A Light-Sensitive Audio Oscillator: The Sonalight *61*

The Sonalight. *61*
Circuit Analysis *62*
Component Description *63*
    555 IC Timer *63*
    The Photoresistor *65*
Construction. *66*
Testing for Operation *67*
Uses of the Sonalight. *67*
    Intrusion Detection *68*
    Observe Shadows or Lights *68*
    Observe Distant Lightning *69*
    Find Archery Target in Dark *69*
    TV Receiver. *70*
    Awaken at Dawn *70*
    Sense Liquid-Level Surface Movement *70*
    Audible Horizontal-Level Device. *70*
    Listening to Your Pulse. *71*
    Finding Sidewalks, Trees and Buildings *72*
    Briefcase *72*
    Police Bubble Machine *72*

7 Polarity-Sensing Continuity Tester Using LEDs

7 Polarity-Sensing Continuity Tester Using LEDs. *73*

Operation. *73*
Circuit Analysis *74*
Testing Bidirectional Devices *76*
    Light Bulbs. *76*
    Transformer. *76*
    Photoresistors *76*
    The Resistor. *77*
Testing Unidirectional Devices *77*
    The Diode *77*
    The Solar Cell *78*
Construction. *79*
Testing for Proper Operation *80*
Additional Uses *80*
A Test Set and Power Supply for LEDs *81*

Visible LEDs. *81*
IR LEDs *81*
Digital Displays *82*

8 A Light-Intensity-Level Wheatstone Bridge *83*

Operation *83*
Circuit Analysis *83*
Construction. *86*
Uses of the Bridge *87*
   Light-Intensity Sensing *87*
   Azimuth Sensing to Light Source. *87*
   Distance to Light Source *87*
   Other Uses. *88*

9 An Audio Continuity and Voltage Tester: the Testone *89*

Background *89*
Operation *89*
   Continuity Testing. *90*
   Voltage Testing *90*
Circuit Analysis *91*
   Continuity *91*
   Voltage Testing *91*
Construction. *92*
Operational Uses *92*
   Bench Tests *93*
   Diodes and LEDs. *94*
   Automobiles. *94*

10 The LED in Motion *97*

DC Operation of the LED *97*
AC Voltage Operation *98*
Bouncing Tennis-Ball Effect *98*
The Rotating LED or SpinLED *100*

LED Light-Bar Module. *101*
Voice Recognition *102*
Stereo Channel Separation *102*
Frequency Measurement and Synchronization. *103*
Digital 1s and 0s *103*
Rotating Machinery. *104*
Vibrating Fiber Optics. *105*

11 Sound Over a Light Beam *107*

The Transmitter. *108*
    The Light Bulb. *108*
    The LED. *108*
    The Laser. *109*
The Modulator *109*
    The Flashlight *109*
    Chopping Light *109*
    The LED. *110*
The Receiver Detector *110*
    The Solar Cell *110*
    The Photocell *110*
    The Photodiode. *110*
    The Phototransistor. *110*
The Audio Amplifier *112*
    The Humbug *112*
    One-Watt Speaker-Amplifier *113*
    Portable Phone Listener *113*
A Simple Light Communications System *113*
Fiber Optics and Light. *114*

12 A Wide-Resistance-Range Audio Continuity Tester *117*

Operation. *117*
    Let Your Ears Do the Walking *118*
    Testing Bidirectional Devices. *118*
Circuit Analysis *119*
    The Unknown Resistance R_x *119*
    Low Current in Test Probes *119*
    555 Operates in Astable Mode.   *119*

Construction.. *120*
Testing for Operation *120*
Additional Uses *121*
    Water Seepage in Boat or Home *121*
    Liquid-Level Height Detector for the Blind........ *121*
    Raindrop Detector *123*

13 Light-Sensitive Detection Device: The Lightstalker *125*

Operation.. *125*
    Initial Setup..................................... *126*
    Color Sensitivity *127*
Circuit Analysis *127*
    Power Supply.................................... *127*
    Output Circuit *128*
    The Photocell *128*
    The Comparator *129*
    Threshold Circuit................................ *129*
    Flicker Filter *130*
Construction.. *130*
    Method of Construction *130*
    Power Transformer.............................. *130*
    Capacitor Quality *131*
Uses for the Lightstalker............................ *132*
    Versatile Light-Change Detector *132*
    Add a Telescope *132*
    Driveway Sensor *132*
    The Sun and Moon *133*
    Fast Moving Objects *133*
    Extended Uses.................................. *133*

14 A Visual Telephone Ringer *135*

"See" The Phone When It Rings..................... *135*
Operation.. *136*
Circuit Analysis *137*
Construction.. *138*
Connection to the Telephone Line *140*
Testing for Operation *140*

15 **The Neon Telelite** *143*

Background of the Telelite *143*
Circuit Analysis *144*
A Loudspeaker Signal *145*

16 **Add Music on Hold to Your Telephone** . . . *147*

Background *147*
Operation *148*
Circuit Analysis *149*
Construction *151*
Connection to the Telephone Line *152*
Testing for Operation *152*

17 **Design Your Own Audio Amplifier Using Transistors** *153*

What We Will Learn *153*
Transistors *154*
    Junction-Type Transistors *154*
    Types of Transistors *155*
    Transistor Test Circuits *155*
    So What's the Difference? *157*
    Beta *157*
A Basic Transistor Amplifier *157*
    Test Setup *157*
    Test Results *159*
    Test Analysis *160*
    Accuracies Expected *161*
Voltage Gain Calculation *162*
More Circuit Analysis *163*
    Static Analysis *163*
    Dynamic Analysis *165*
Input and Output Impedance *165*
    Background *166*
    Output Impedance *166*
    Input Impedance *167*
Transistor Amplifier Design *167*
    The Steps *168*

Examples of the Design Process *171*
   One-Stage Audio Amplifier *171*
   Two-Stage Audio Amplifier *172*

Preface

Are you tired of the old bell-ringing sound of the telephone? Then, build a musical solid-state telephone bell with a volume control so you can use it on the patio without disturbing the neighbors. You can assemble a simple test instrument using the versatile light emitting diode that will tell you if a voltage is ac or dc. If you want to find your way to a light source, the light compass in Chapter 8 will do this for you. How would you like to play music over a light beam and experiment with fiber-optic light pipes. This, also you can do.

When you build the light-sensitive audio oscillator you will be able to locate a bull's eye in the dark with your eyes closed. Did a power failure at night make you miss your flight? Build an audio alarm that will awaken you during a power failure. And how about an audio continuity tester that lets you do the testing using your ears instead of your eyes? Apply an audio voltage to an LED and move it back and forth rapidly, and a new kind of visual display will be revealed to you.

You'll build a continuity tester that produces a tone that changes with resistance so that you can check for a short circuit (which produces a tone) or a 30-megohm resistance (which gives a "plop-plop" signal). Another project that you will construct is a tone alarm that signals you when the liquid in a pot on the stove is boiling. You can detect a person breaking a beam of light over 100-feet away using a simple passive light detector that avails itself of nearby light sources such as playgrounds or street lights. And if you are tired of the ringing of the telephone bell, use a device to turn on a table lamp or spotlight when the phone rings. You can be cutting the grass in the backyard and still not miss any important calls due to noise from the mower. The music-on-hold feature that you can add to your phone will entertain your callers with music or commentary while you go to another phone to talk to them. You won't have to walk back to hang up the other phone. The chapter on transistors will help you understand how they operate so that you can design your own audio amplifier. This simple approach and understanding will make you feel good about a difficult subject.

You will enjoy building these simple and inexpensive projects, you'll improve your electronic skills, and you'll have some interesting devices to show your friends, neighbors, and schoolmates. Students might even find the subject for an entry into a science fair project that will start you on a new career in electronics or science. May you have many warm nights in the winter and cool evenings in the summer as you build these one-evening electronic projects.

CALVIN R. GRAF
RICHARD S. GOSS

Acknowledgments

As with previous books, grateful acknowledgment is due Forrest M. Mims III, a prolific technical author and innovator in optoelectronics, technical presentations, and encouragement to others to excel in any technical endeavor. His many unique observations of our technical surroundings and his easy-to-understand writings serve as a lasting inspiration to the technically-minded youth of our country. Dr. Guido Merkens, Concordia Lutheran Church, San Antonio, for over a quarter century, has continued to shine the light that reminds us that all things are possible when the Spirit is right and that there is "good measure, pressed down, and shaken together, and running over." All photographs, except where otherwise indicated, were taken by Larry W. Graf. Photoprocessing and copying, except where otherwise indicated, are by Ricky Martinez. Grateful thanks are due to the various manufacturers and electronic suppliers for their very helpful efforts in providing photographs. Finally, acknowledgement is due our wives, Edla (CRG), and Lin (RSG), who for many years have listened patiently, but perhaps with some apprehension, to the strange sounds that radiated from some of the projects on the workbench. We know those beeps and chirps did not really overcome!

1 Construction Assembly Hints

This book describes in detail a number of electronic projects which can be constructed and assembled in one evening by a newcomer to the fascinating world of electronics. The projects to be described also could possibly fall in the category of the unique and innovative, even to the skilled or experienced electronic project builder. Some of the projects might elicit, "Oh, I knew *that*," but building a simple instrument that will aid you greatly almost every time you need and use it is vastly different from just wishing for something. When you can make use of a simple light-emitting-diode (LED) test device that will give you as much information as a several-hundred-dollar cathode-ray-tube (crt) oscilloscope, you will realize that simplicity and functional use do have their place in electronics. Let's take a general look at the projects we will cover in the chapters to follow.

Background of Projects

In each of the projects the device is described in detail. This includes the technical characteristics of the associated electronic components, some known operational uses (as you learn to innovate, you will think of many uses yourself), and the technical operation of the circuit. Some areas are suggested for further exploration for uses of the device which you can undertake as your time and interests permit. The beginning student of electronics, the experimenter, electronics hobbyist, science fair entrant, science teacher, and practice-oriented builder will all find this book of interest. Most of the projects are electrically simple, employing a small number of parts so that the devices can be put together inexpensively and without a lot of electronic-circuit knowledge. Thus when you assemble one of the projects and it works first time off, you can proudly say, "Hey, look, it works—just as though I knew what I was doing!"

The Probevolt

This is the simplest of electronic test devices that gives you a lot of information about voltage in a circuit, battery condition, and polarity of the associated voltages. It will operate over a voltage range of about 1 to 28 volts ac or dc. The *Probevolt* consists of basically three components: two LEDs (light-emitting diodes) and a resistor. It can be assembled in a used felt-tip pen or marker. It is small and very handy, as it is handheld. The Probevolt is covered in Chapter 2.

LED Voltage and Polarity Indicator

This device is very similar to the Probevolt but is designed to cover a greater voltage range through use of a switch and a number of different resistor values. It will operate over a voltage range of 2 to 120 volts ac or dc and is packaged in a small plastic case. This device is covered in Chapter 3.

A Solid-State Telephone Bell

The "ring" of a telephone bell is what telephone users have become used to for over 100 years. It is only recently that phone "ringers" have become a slightly musical note rather than the clanging sound of an electromechanical bell. The *Sonabell* is described in Chapter 4.

The Poweralert

Have you ever had an electrical power failure at night when you were asleep and then missed an important appointment because your electric alarm clock did not go off on time? Well, you can avoid many problems by building the *Poweralert*, which will sound an alarm when the power goes off and stays off. You can then set a manual alarm clock to wake you up on time. Details are given in Chapter 5.

A Light-Sensitive Audio Oscillator

The *Sonalight* is an audio oscillator that changes frequency with the amount of light reaching a photocell. The more light present, the higher is the frequency. The lower the light level is, the lower is the frequency. It is so sensitive to small amounts of light that it can almost tell the difference between starlight and no starlight. Many useful applications of the Sonalight are described in Chapter 6.

Polarity-Sensing Continuity Tester Using LEDs

Once again the popular and extremely useful LED is put to service in a circuit that checks continuity and any associated polarity of items such as light bulbs, diodes, LEDs, hardwire circuits, motors, buzzers, resistors, fuses, etc. This handy aid is covered in Chapter 7.

A Light-Intensity-Level Wheatstone Bridge

Use is made of a very old circuit—the Wheatstone bridge—employing a photocell, LEDs, and a battery to determine when certain light intensity levels are balanced. With this light bridge you will be able to tell when you are aiming an object to the left or right of a light source. Further uses are described in Chapter 8.

An Audio Continuity and Voltage Tester

This test device produces a tone when there is continuity in the circuit being tested. Because a tone is produced when there is continuity, it is not necessary to take your eyes off the test points as is necessary with an ohmmeter. Items such as lamps, resistors, etc., can be easily tested for continuity. In the voltage test mode, a tone is produced to show the presence of voltage levels from 1 to 28 volts, ac or dc. This device is especially useful when you are working on automobiles, and it is fully described in Chapter 9.

The LED in Motion

We tend to think of an LED, neon bulb, or light bulb as being either *on* or *off*. Using an LED we will see that when we apply an alternating voltage, such as 60 Hz or audio output from a stereo amplifier, to such a device that it can be seen to turn on and off, but only if we move it rapidly enough. This new form of aural-visual entertainment and test instrumentation device is covered in Chapter 10.

Talking Over a Light beam

Since the ancient days of the Greeks and Romans, light has been used as a means of communications. The light, however, was used for signaling only ("one if by land, two if by sea") and talking over the beam of light produced by a lamp, bonfire, or flashlight did not occur. In Chapter 11 we will discuss talking or sending music over light beams produced by LEDs and flashlights. The light beam can pass through space or a fiber optic strand or cable.

Audio Continuity Tester for Wide Resistance Range

An adaptation of the circuit used with the light-sensitive audio oscillator permits us to make an audio continuity tester that can be used over a wide resistance range. With this circuit we will be able to hear the tone produced by a short circuit (0 ohms) and an almost open circuit (30 megohms). It is covered in Chapter 12.

The Lightstalker

This novel circuit is a light sensor that uses available light to serve as an intruder detector. The sensitive device adjusts itself to light from a nearby incandescent source such as a street light, porch light, or perimeter light. When the light level is broken (or reduced by a passing intruder or passerby) the unit sounds off at a low level. It adjusts automatically to slow changes in light such as the rising and setting sun and moon. It will detect a man breaking a beam of light over 100 feet away. The *Lightstalker* is described in Chapter 13.

A Visual Telephone Ringer

There are times when you don't want to hear the phone ring and there are times when you can't hear the phone ring. When the phone rings and there is a lot of noise or music in the room or office, this device lights an ordinary lamp that is plugged into it. Perhaps the baby is asleep and you've unplugged the phone or turned off the ringer. You won't miss any calls because the circuit described in Chapter 14 will light a silent "light" bell for you.

The Neon Telelite

When you don't need a light telephone ringer as "big" as the visual telephone ringer, the *Telelite* is just the device for you. A neon lamp is made to turn on when the phone "rings" such as when you might have on your stereo earphones and outside sounds are not easy to hear. The Telelite is used in sound studios to indicate telephone ringing without the sound. The Telelite is covered in Chapter 15.

Add Music on Hold to Your Telephone

You can be like the big boys by providing music when someone calls you and you place them on hold, either to go to another instrument in a different room, or to call another person to the phone. This handy circuit is described in Chapter 16.

Design Your Own Transistor Audio Amplifier

Transistors buffalo you still? This simple and easy-to-understand chapter on transistors will remove much of the mystery of how they work. When you finish, you will be able to design your own audio amplifier using readily available transistors and parts. Chapter 17 covers operation of the transistor amplifier in simple and easy-to-understand detail.

Circuit Diagram Electronics Symbols

We are all used to reading roadmaps and know how we must follow instructions closely in order to leave one city and drive to another, arriving there without too many deviations or lost time. In the same manner a circuit diagram is used to show how the components in a circuit are connected in order to produce the desired result, such as an audio oscillator, a test device, and the like. Since most of the symbols follow an international plan a radio made in other countries has a circuit diagram that can be fully read and understood in the U.S. This is especially important when you attempt to repair or maintain a piece of electronic equipment with which you are not familiar. The electronics symbols are essentially an international language. A circuit diagram is also known as a *schematic*, which means that the circuit follows a scheme. Some of the most widely used schematic symbols are shown in Fig. 1-1. Most of the symbols depict what they do electrically, such as an LED emitting light with arrows, or how they are made up physically, such as the parallel plates of a capacitor. There is usually good reasoning behind the development of the symbols as new devices are invented and brought into the electronics inventory of circuits. The symbol for the fixed resistor (a zig-zag line) progresses to a variable resistor (such as a volume control) by adding an arrow to show that the value of resistance is a variable one. The diode symbol becomes the symbol for an LED, which emits light, by adding arrows depicting the emission of light. Three symbols are combined into one in the photoresistor symbol, which shows a resistor that changes resistance value when incoming light strikes it (resistance decreases with increase in light).

A device that has been around for a while but which does not have a symbol is the piezoelectric sounder, buzzer, or annunciator. It is all solid-state and has no moving parts. A suggested symbol for the piezoelectric sounder (Fig. 1-1) is derived from that used for a signal generator—a wavy line or sine wave. In our new symbol we will use two sine waves, one above the other. The input to the piezoelectric sounder is a dc voltage and the output of the piezoelectric sounder is almost a pure tone of sound energy.

Component Values

Most components are marked with a number to indicate what they are and what they do. A diode might be a 1N34, a transistor 2N108, a tv picture tube 25DRP4, and an integrated-circuit chip 555. Capacitors are usually marked with their value in microfarads, such as .01, .001, or 100, and percentage of tolerance.

Resistor Color Coding

Resistors used in electronic circuits can vary greatly in resistance value and are coded with a color code to show their value and percentage of tolerance. Fig. 1-2 shows the color code for resistors.

Since a resistor is one of the harder items to identify as to its resistance value, let's discuss what we are to do. The average resistor used in electronic circuits is the $1/2$-watt size (about $1/2$ inch long) and this is usually the size that is sold at the local radio stores. The resistors can be bought in card packs of two or in assorted values of 100 resistors in a plastic box. If we pull one out of the box and see that the colors are red, red, and red, we can say 2, 2, and 2 zeros (multiplier of 100), for 2200 ohms, where an "ohm" is the basic unit of resistance. If the colors are brown, black, orange, we say 1, 0, and 3 zeros (multiplier of 1000), for 10,000 ohms, or 10 kΩ, where k stands for 1000 and Ω is the Greek letter Omega and stands for ohms. A 68-ohm resistor would have a color code of blue (6), gray (8), and black (multiplier of 1) for 68 times 1, or 68 ohms. An interesting value is a 1-ohm resistor where we have brown (1), black (0), and gold (multiplier of 0.1) for a value of 10 times 0.1, or 1 ohm. You can work a long time in electronics and not find a circuit that calls for a carbon resistor of 1 ohm! Wirewound resistors are usually used for those low values of resistance. After you have worked with resistors for a while you will recognize that if you want a 10 kΩ resistor you should look for orange on the third band (000), disregarding all other third-band colors because their value will be too high or too low. When you sort out an orange third band (000), then look at the other colors to see if they are brown (1) and black (0), for 10 kΩ. Note that if there is no fourth band for tolerance, the tolerance is plus or minus 20 percent. A 100-ohm resistor could be measured as 80 to 120

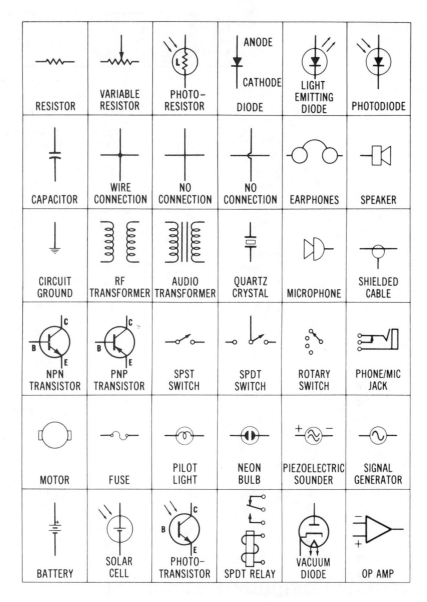

Fig. 1-1. Electronic circuit symbols.

ohms and still qualify as a 100-ohm resistor. In almost all circuits the 20-percent tolerance will be satisfactory for circuit operation. As you would expect, the closer the tolerance (10, 5, 1 percent), the more the resistors cost. A resistor color-code guide is shown in Fig. 1-3. This handy guide uses thumb-wheels to set up colors, which gives you the resistance value.

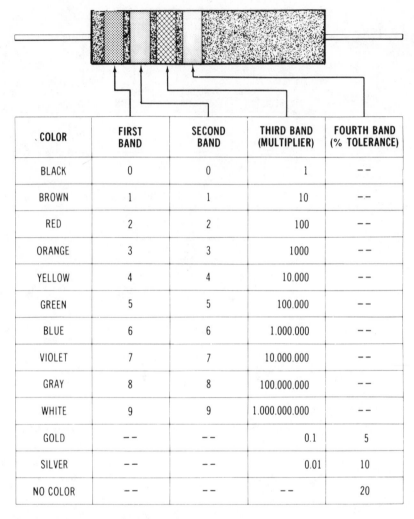

COLOR	FIRST BAND	SECOND BAND	THIRD BAND (MULTIPLIER)	FOURTH BAND (% TOLERANCE)
BLACK	0	0	1	– –
BROWN	1	1	10	– –
RED	2	2	100	– –
ORANGE	3	3	1000	– –
YELLOW	4	4	10.000	– –
GREEN	5	5	100.000	– –
BLUE	6	6	1.000.000	– –
VIOLET	7	7	10.000.000	– –
GRAY	8	8	100.000.000	– –
WHITE	9	9	1.000.000.000	– –
GOLD	– –	– –	0.1	5
SILVER	– –	– –	0.01	10
NO COLOR	– –	– –	– –	20

Fig. 1-2. Color coding for resistors.

Electronic Component Mounting

Electronics construction and assembly techniques have made great strides since the invention of the transistor and integrated-circuit chip. A large number of components can be mounted on circuit boards that take just hundredths or thousandths of the space that vacuum-tube components took years ago. With the coming of the IC chip came the use of the word *discrete component*, which is used to describe separate transistors, resistors, capacitors, and the like, when used in a circuit.

"Integrated circuit" (IC) means that all the components (transistors, resistors, etc.) are integrated and built into layers of a chip. An IC chip the size of the pupil of your eye can consist of thousands of transistors, diodes, and resistors, all of them consuming perhaps just a few millionths of a watt of power. A good example is the quartz liquid-crystal–display wristwatch whose battery can last from two to five years. That is, indeed, an electronics breakthrough.

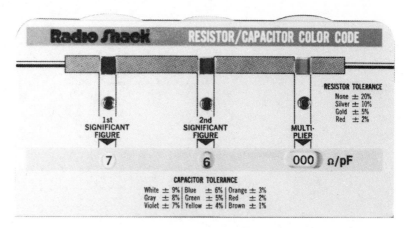

Fig. 1-3. Resistor color code guide. *Courtesy Radio Shack, Div. of Tandy Corp.*

Breadboard

But things were not always that sophisticated. Back when there weren't many local neighborhood radio parts houses it took days to design and lay out a simple circuit. After the circuit was designed on paper the parts were laid out and actually placed on a breadboard (perhaps borrowed from mother's kitchen!). The parts could be arranged to best suit the operation and practical aspects of the circuit. Components were mounted in a temporary manner and the circuit might even be made to work to show that it was practical. With this reassurance the laborious task of drilling holes, mounting vacuum-tube sockets, and mounting support strips to hold resistors and capacitors began. It was a slow and tedious job.

Perforated Boards

There are a number of different sizes of perforated phenolic circuit boards which can be bought at the local radio and electronics parts houses. These boards are prepunched with standard sizes of holes spaced center-to-center. Fig. 1-4 shows two of the various sizes which are available in different lengths and widths. These are easily cut into the desired size of board. The perforated boards are also known as *perfboards* and *P-boards*. An IC chip will mount directly into a board with 0.1 inch (0.25 cm) spacing. This is the spacing between pins on all IC chips, whether they are 8-pin, 14-pin, or 16-pin DIPs (dual in-line plastic). As the complexity of the IC chip increases, the number of pins needed to input and output the data increases. Table 1-1 shows some common spacing and hole dimensions on various perfboards.

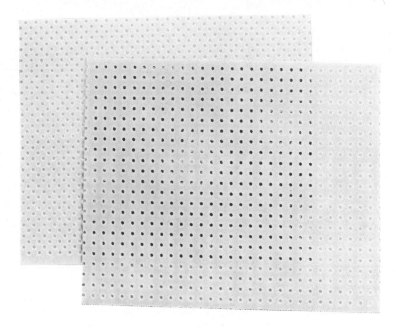

Fig. 1-4. Various sizes of perforated circuit boards. *Courtesy Radio Shack, Div. of Tandy Corp.*

Table 1-1. Perforated Prepunched Circuit-Board Sizes

Hole Spacing		Hole Size	
Inches	**Centimeters**	**Inches**	**Centimeters**
0.100 × 0.100	0.254 × 0.254	0.047 ($^3/_{64}$)	0.119
0.143 × 0.143	0.363 × 0.363	0.0625 ($^1/_{16}$)	0.158
0.1875 × 0.1875	0.476 × 0.476	0.0625 ($^1/_{16}$)	0.158
0.265 × 0.265	0.673 × 0.673	0.093 ($^3/_{32}$)	0.236

Printed-Circuit (PC) Board

The use of the printed-circuit board was made popular as a result of the transistor, which operates with low values of current. As a result point-to-point wiring is not necessary because the thin copper foil etched on the board is sufficiently "large" to handle the current required for most transistor operation. When the IC chip came along it operated with even less current and blended in beautifully with the pc board concept, even eliminating the IC chip socket holder (necessary for certain applications). Fig. 1-5 shows a dual in-line IC pc card which will accept a 14-pin or 16-pin DIP chip or two 8-pin chips. The 16-pin holder can be carefully sawed in half so that it can be used with the IC chips we will cover in Chapters 6 and 12. The IC DIP pins are inserted from the side opposite the copper foil, and the pins soldered to the foil. Further connections are then soldered to the large foil areas as required.

Solderless IC Printed-Circuit Board

An experimenter's aid available from many electronic parts houses consists of the same pc card shown in Fig. 1-5, an IC socket into which different IC chips can be easily inserted, and 16 spring terminals. The IC socket is inserted from the side opposite the foil and the 16 pins are carefully soldered to the copper foil. The 16 spring clips are inserted in the predrilled holes and soldered from the foil side. It is then very easy to push down the spring clips while hookup wire is inserted. This unit can be used over and over for different circuits. Fig. 1-6 shows the spring terminal. When you are firm on a circuit you may then want to go to the pc card (Fig. 1-5).

Cardboard

Cardboard from two-sided cardboard packing boxes can be used for mounting-experimental circuits. Cut up a sturdy box into pieces of cardboard 3 × 3 inches (7.62 × 7.62 cm) or 3 × 6 inches (7.62 × 15.24 cm). Mount the resistors, capacitors, diodes, LEDs, photocells, and the like, by using a pair of long-nose pliers to force the wire lead through the cardboard. You can also use a small nail, stiff wire, ice pick, or scribe to make small holes in the cardboard through which the copper leads are inserted. Bend the wires over on the bottom side and make any additional connections on the same side of the cardboard. This means of mounting electronic components will let you get started on a project until you can obtain more permanent means of mounting.

Wiring and Soldering Methods

Wiring and soldering techniques have changed considerably in the past 30 years. This was, again, brought about primarily by the invention of the transistor and the evolution of the pc board and the IC chip. Since these components require less power to operate, they are more delicate and much smaller than their predecessors.

Fig. 1-5. Printed-circuit card with IC chip. *Courtesy Radio Shack, Div. of Tandy Corp.*

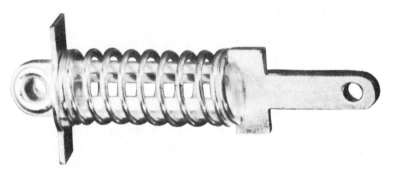

Fig. 1-6. Solderless spring terminal for pc board. *Courtesy Radio Shack, Div. of Tandy Corp.*

Point-to-Point Wiring

Laying out a pc card so that all the components can be mounted and yet be interconnected by the copper foil is indeed an art in itself. For experimental circuit design, where the layout is apt to change, or for maintenance or design purposes, discrete point-to-point wiring is used. As the name implies, a piece of wire is run from one point to another point as might be required by the circuit. Since insulated wire is used, a number of wires can be run alongside each other. Point-to-point wiring is used by the various telephone companies using a wire-wrap technique. A special tool is used to wrap the wire around a terminal a number of times. The electrical connection is made by mechanical means only. There is no soldering involved. Fig. 1-7 shows a manual wire-wrap tool. The technique is extremely rapid and connections can be changed merely by unwrapping the wire. Point-to-point wiring is the method we will use in most of the projects described in the chapters to follow.

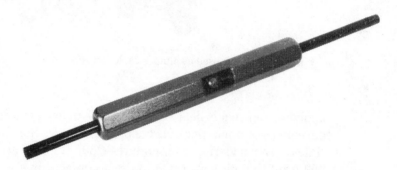

Fig. 1-7. Point-to-point wire-wrap tool. *Courtesy Radio Shack, Div. of Tandy Corp.*

Test Clip Wiring

When experimenting with various circuits it is very handy to have several sets of color-coded test cables. These cables are 10 to 14 inches long and are ready-made with small alligator clips on each end so that you can clip to speaker terminals, LEDs, resistors, and the like. They can be used to allow you to test a circuit rapidly for proper operation before you hardwire it. You will find them most useful. Fig. 1-8 shows the clips in a package. The various colors of wires will allow you to use red cables for connecting to a positive voltage source such as a 9-volt transistor battery, a black cable for the ground connection, yellow for connection to the anode of an LED, and green for connection to the cathode of the LED. Test the clips for continuity from time to time to be sure they are good, especially when you use them a lot.

The Soldering Iron

A small, low-power, soldering iron must be used when working with transistors, IC chips, LEDs, and pc cards. Some standard soldering-iron wattages available are 15, 25, 30, and 40 watts, at 120 volts, ac or dc. If you use too heavy a soldering iron, you may damage the delicate components by

Fig. 1-8. Test clips used as jumper cables for quick wiring. *Courtesy Radio Shack, Div. of Tandy Corp.*

applying too much heat. Consequently many devices show the heat rise permitted for a number of seconds. Many LEDs thus have a precaution that states: "lead soldering temperature—230° C for 7 seconds." Ordinarily you will want to limit your soldering-iron contact time with an LED, transistor, or IC chip to less than 10 seconds. So wait a short while before reapplying soldering-iron heat if a second application is necessary.

Cleaning the Iron

A soldering-iron tip easily becomes corroded after use, so it is necessary to clean the tip before using it to solder. After the iron has warmed up, clean it on a damp sponge, wipe it with a clean cloth, and, if necessary, stroke it a few times across a piece of sandpaper to clean it.

Tinning the Iron

Before using your iron to solder, tin the tip by applying solder to the tip so that it spreads evenly on all sides. Be sure to use rosin-core solder that is made of 60 percent tin and 40 percent lead. Remember, *do not* use acid-core solder for any kind of work with electronic equipment; it is extremely corrosive and can ruin a perfectly good circuit board. Save the acid-core solder and heavy soldering iron for gutters and other sheet-metal work.

Soldering Wires and Components

Soldering electronic parts can be a lot of fun because there is skill involved. If you are new to the use of a soldering iron and solder, practice a while on some old, discarded parts so that you won't ruin anything that you want to keep. For a connection to a terminal cut the insulation back to ½ inch on the connecting wire, and, using a pair of long-nosed pliers, bend the wire back around the terminal so there is some mechanical support. Touch the heated soldering iron to the terminal and wire and, after a few seconds, touch the solder to the terminal and the solder will flow freely around the wire and terminal. Immediately remove the solder and iron. You can lightly blow on the connection to aid cooling, but do not move the connection for at least 10 seconds. This will ensure proper cooling of the solder. If you do not apply enough heat to melt the solder properly, or move the wire before the solder has cooled, you might create a defective or "cold solder" joint. These joints are difficult to locate when troubleshooting a circuit that does not operate properly, because to the eye the joint appears to be perfectly good, but there is no, or a very-high resistance, connection to the terminal.

Soldering IC Chips and LEDs

As we discussed earlier under the section of the wattage size of the soldering iron, IC chips, LEDs, diodes, and transistors are subject to damage by heat. Therefore, when you are soldering these devices, use the minimum amount of soldering time possible. The leads to an IC chip solder very easily to the copper foil of a pc board, but use the least amount of solder that will allow a good connection to be made. You will feel especially good after you have completed the job of soldering all the components into a handy, working project.

Unsoldering Components

Almost all electronic components can be used over and over again if care is taken in unsoldering them from terminals, connections, and other components. Again, use the minimum amount of heat that will melt the solder while removing the leads of the component from the terminal.

The Desoldering Tool

There are a number of desoldering tools available which are used to suck up melted solder from an electrical connection after it has been melted by a soldering iron. The tool can really ease the task of unsoldering components without damaging them. Fig. 1-9 shows a tool which is spring operated to create a vacuum and suck up the melted solder. Others use a rubber "squeeze" bulb.

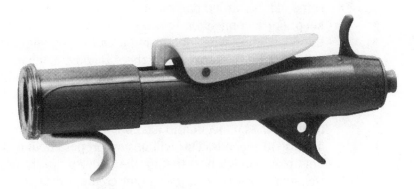

Fig. 1-9. Spring-loaded desoldering tool. *Courtesy Radio Shack, Div. of Tandy Corp.*

The Solder Wick Absorber

A solder wick absorber is used to draw up or suck up melted solder from an electrical connection after it has been heated by a soldering iron. The remover wick is available in rolls at the various radio supply houses. Pieces of the wick are cut off and discarded after they have absorbed all the solder they can hold. The wick absorber sucks up solder in almost the same way the paper towels absorb spills on the tv commercials!

Packaging the Project

There are a number of different ways in which your project can be packaged. We will discuss these in greater detail in the chapters that describe each project.

Metal Cases

Metal cases, boxes, and cabinets are available at most electronic supply houses. Many have small rubber feet, removable covers, and two-tone colors. The price will vary from under $2.00 to nearly $10.00 for the size to fit the projects discussed.

Plastic Cases

These are often called "experimenter boxes," have an aluminum cover, and vary in size form a little over 3 inches (7.62 cm) to almost 8 inches (20 cm) in their largest dimension. Their price is most attractive for kit building and varies from a little over $1.00 to under $3.00. You can also use any plastic case you might have handy around the house, varying from a small pillbox, a toy car container, to the round top cover of a spray can.

Felt-Tip Marker

A used felt-tip marker is very handy to use as a package for the simpler projects. Simply cut off the tip and remove the dry ink filler.

Bandage Boxes

Small bandage-container boxes, metal-cover tops from spray cans, and any assorted small metal cases make ideal cases in which to mount your project. You can use any size or shape that fits your requirement, metal or plastic.

Lettering Your Project

There are a number of hand-operated lettering machines available at your neighborhood discount or stationery store, and also at most electronics stores. These can be used to very neatly and professionally label the project as well as the various operating controls, jacks, and the like. You will find that a nice label or directions can really help dress up a project and show the final touch of a job well done.

2 The Probevolt Voltage Detector

The Probevolt is a passive voltage-level indicator with an autopolarity feature. The circuit will work on alternating current (ac) voltages or direct current (dc) voltages of from 1 to 50 volts of either polarity without the necessity or reversing the test leads to the circuit under test. Fig. 2-1 shows the Probevolt packaged in a used felt-tip pen.

Much as a logic probe allows the user to view the static logic states of a circuit module or integrated circuit, the Probevolt indicates the presence of a voltage, its polarity, whether it is ac or dc, and its relative voltage level. The Probevolt also works on square waves of any polarity, sawtooth voltages, or any unusual voltage shape.

Circuit Analysis

The circuit employed in the Probevolt is very simple, using two light-emitting diodes (LEDs) connected in reverse-polarity parallel, with a common series current-limiting (or voltage-dropping) resistor. Fig. 2-2 shows the circuit with the anode of an LED, X1, connected to the tip of the probe. When the tip is touched to a terminal with a positive voltage present, X1 will light. At this time the negative clip would be connected to circuit ground or a negative voltage terminal of the circuit to be tested.

When we trace electron flow or movement in the probe circuit we see that electrons enter the negative (clip) terminal, travel to the cathode of X1, across to the anode of X1, producing light at that time, through the 1-kilohm

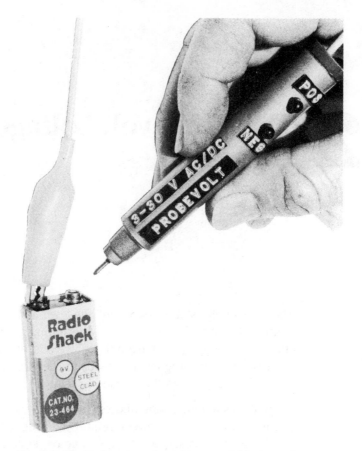

Fig. 2-1. The Probevolt packaged in a used felt-tip marker pen.

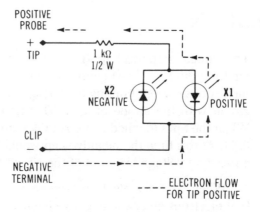

Fig. 2-2. The LEDs of the Probevolt vary in brilliance with the level of input voltage.

current-limiting resistor to the positive terminal of the tip and voltage source. However, at this time, electrons cannot proceed from the anode of X2 to the cathode of X2 and so X2 does not light. When the tip is connected to a negative voltage, and the clip to a positive voltage, electrons will flow in the opposite direction and X2 will be lit while X1 will be out. In this way we can use LEDs to tell us what the polarity of voltage is at the tip of the probe.

The Light-Emitting Diode

In order to fully appreciate the marvels of the LED, let's first take a look at the diode—a two-element electronic device.

The Diode

An ordinary diode, be it a solid-state diode or vacuum-tube rectifier, is used to rectify or change an alternating current (ac) voltage into a pulsating direct current (dc) voltage or to act as a switch which permits only the correct polarity of current to flow. Fig. 2-3 shows the action that takes place in the diode circuit. These devices usually are used in circuits with ac coming in and dc going out, whether the voltage is audio (20 to 18,000 Hz) or radio frequency (10 kHz to 1000 GHz). At this time the diode is rectifying or changing the current. No other action is taking place.

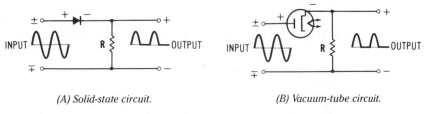

(A) Solid-state circuit.   *(B) Vacuum-tube circuit.*

Fig. 2-3. Rectifying action in a diode circuit.

The LED

The LED, however, ordinarily has dc, or pulsating dc, applied to it. Because of the low back-voltage, or reverse-polarity voltage (3 to 5 volts) of the LED it cannot be used on ac unless the LED has a diode or LED connected in parallel as shown in Fig. 2-4. When LEDs are operated as pilot lights, status indicators, current intensity indicators, and digital number segments, they are connected as shown in Fig. 2-5. The usual current-limiting resistor is always used with an LED so that it does not draw excessive current, which reduces its lifetime.

How the LED Produces Light

Let's take a closer look at the "innards" of an LED to see how it produces light. In Fig. 2-6 we see the basic building blocks of LEDs, diodes, transistors,

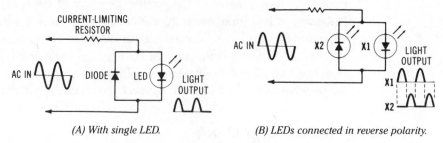

(A) With single LED.   *(B) LEDs connected in reverse polarity.*

Fig. 2-4. Parallel diode/LED operation.

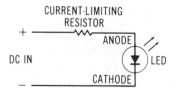

Fig. 2-5. In dc operation of an LED current polarity must be observed.

and other solid-state devices. In this simple circuit arrangement we have omitted the series current-limiting resistor. The n-type semiconductor material has excess numbers of electrons, hence the *n* designation for *n*egative. The p-type semiconductor material has an absence of electrons, or it has an excess of holes, hence the *p* designation for *p*ositive.

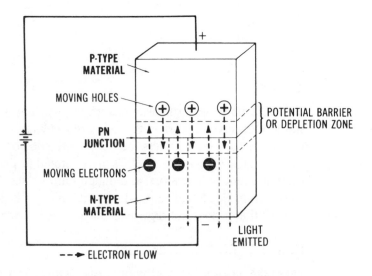

Fig. 2-6. Forward-biased pn-junction LED emits light.

When a current flows through the pn junction (where the p material and n material meet) both holes (absence of electrons) and electrons participate in current across the pn junction to fill holes, and the holes move across the junction to occupy spaces (holes) vacated by the electrons. Light is generated when an excited electron returns to its normal state of equilibrium by combining with a hole in the valence band—its state of rest in the atomic structure. The junction (pn) diode is the device for raising a number of electrons into the excited state so that they can fall back into a state of equilibrium and produce light while doing it. After the junction has been crossed the electrons fall into holes on the p side and in the process emit photons of light. Gallium or phosphorus is ordinarily used in the manufacture of LEDs. The n side of an LED junction absorbs much less light than the p side, so the n side is usually employed as the main light-emitting region. Thus LEDs are made so that the light generated at the pn junction has to travel just a short distance before being emitted into space.

When a pn junction is reversed biased (Fig. 2-7), there is no electron flow or hole movement because the negative terminal connected to the p material repels the electrons and the positive polarity of the n material repels the holes. Therefore no light is emitted.

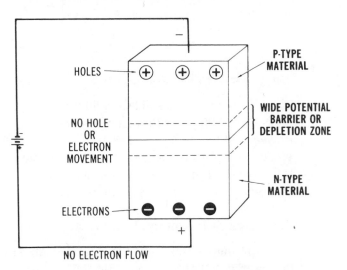

Fig. 2-7. Reverse-biased pn-junction LED emits no light.

By referring to Fig. 2-8 we can see a plot of junction current against applied voltage and light output from the LED. Note how constant the voltage is at 1.6 volts for the red LED. Light output increases directly with current until heating occurs. Red LEDs (655-nanometer wavelength) are made of gallium arsenide (GaAsP) with a constant voltage drop of 1.6 volts dc. Green LEDs (550-nanometer wavelength) are made of gallium phosphide (GaP) with a constant voltage drop of 2.0 volts dc.

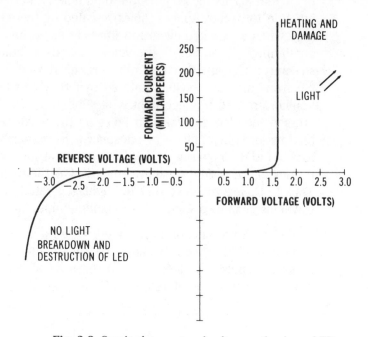

Fig. 2-8. Graph of current and voltage action in an LED.

The LED: A Solid-State Light

The LED is, indeed, a solid-state light and a miracle in its own right. Its half-life is about 100 years (brilliance reduced to half), it doesn't have a filament to wear out by giving up electrons as does a light bulb, and it is so rugged you can drop it on a sidewalk without it shattering. The LED can take ten times overcurrent (up to several hundred milliamperes) without it burning out, though its lifetime would be reduced slightly.

Brilliance Tells Voltage Level

The light output from an LED increases directly with current through the device. Fig. 2-9 shows how light output begins the instant the junction is forward biased and continues to increase directly with current increase until several hundred milliamperes are reached. At this time heating and damage begin to occur and the light output begins to drop off.

If you hold an LED close to your eye you can see the pn junction glowing, with only microamperes of current flowing. In fact, using this technique you will be able to see water and dust particles floating on the surface of your eye. And if you use a small magnifying glass of 5 to 10 power

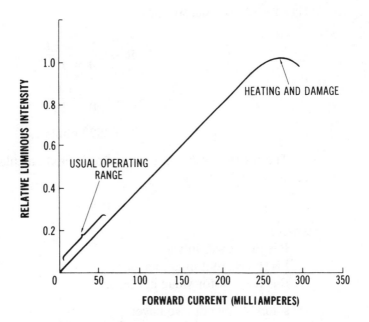

Fig. 2-9. Typical luminous intensity versus forward current of LED.

you will be able to see the internal construction of the LED along with its cat's whisker connection to the pn material. Be sure to use a clear, non-diffusing lens for these observations. The LED is really marvelous in its construction, simplicity, usefulness, and production of "cool light."

Current-Limiting Resistor

The typical LED operates at 20 mA of current with rated light output, though there are now LEDs with currents of 5 mA at rated light output. A Probevolt can be designed to operate from low voltages of 1 to 2 volts up to 30 to 40 volts. At a few milliamperes of current the light output will be low, while at the higher currents (voltages) it will be rather bright. If you want nominal brilliance of the LED at 12 volts, you can calculate the value of the series dropping resistor by using the formula

$$R = \frac{E_{in} - E_{LED}}{I}$$

where,
    R is the resistor value in ohms,
    E_{in} is the Probevolt input voltage in volts,
    E_{LED} is 1.6 volts for red LED, 2.0 volts for green LED,
    I is current in amperes, chosen as 20 mA (20×10^{-3} ampere).

Thus for 12 volts dc, we have:

$$R = \frac{12 - 1.6}{20 \times 10^{-3}}$$

$$= \frac{10.4}{20 \times 10^{-3}}$$

$$= 520 \text{ ohms}$$

The wattage value of the resistor is next calculated using the following power formula:

$$P = I^2R$$

where,
    P is the power in watts,
    I is the current in amperes,
    R is the resistor value in ohms.

Thus, for 12 volts dc, we have:

$$P = (20 \times 10^{-3})^2 \times 520 = 0.208 \text{ watt}$$

We would use a $\frac{1}{2}$-watt resistor as this is a standard and is readily available.

Construction

The Probevolt is a hand-held instrument, and as such it is best assembled in a small volume.

The Case

The Probevolt is contained in a felt-tip marker pen case. Use a pen that has dried out and cut off the end about $\frac{1}{2}$ inch down. Save the colored insert because you will use it again. Remove the felt tip as well as the ink-retaining material in the body of the pen.

Parts List

Fig. 2-10 shows the parts which we will assemble and insert into the body of the empty pen. Table 2-1 is a parts list. All of these items are available at the local radio electronics supply and parts houses. You can buy packaged test clip leads of various colors; it will be useful for you to pick up a package of these as you will use them for other projects in the chapters that follow.

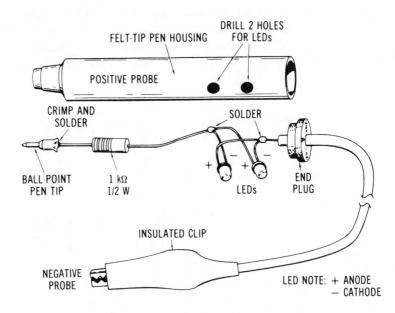

Fig. 2-10. Assembly of components for Probevolt.

Table 2-1. Parts List for Probevolt

Item	Description
LEDs	Two Red, or Green LEDs
Resistor	1-kΩ, ½-W
Test Lead	24 to 36 inches (60 to 90 cm), with Alligator Clips
Miscellaneous	Felt-Tip Pen, Ball-Point Pen Tip, Solder, Soldering Iron

Tip of Probevolt

The tip of the Probevolt is a used ball-point pen tip cut off several inches from the tip. Be sure to remove the ball from the tip or it may provide a poor tip connection for the probe. You will not be able to solder it in place since it is usually made of steel. Solder one lead of the resistor into the brass portion of the tip. Refer to the sketch in Fig. 2-10 for further details on the tip of the Probevolt.

Assembly

Locate and drill two holes in the body of the probe in which the two LEDs will be inserted. Position the components as shown in Fig. 2-11 and solder leads where required. Note that a hole is drilled in the top end of the felt-tip pen and the negative lead is inserted through the end plug before soldering and sliding all components in the probe housing. You should also tie a strain-relief type of knot in the end of the flexible lead in the housing. Use epoxy or a good glue to hold the probe tip rigid in the tip of the probe.

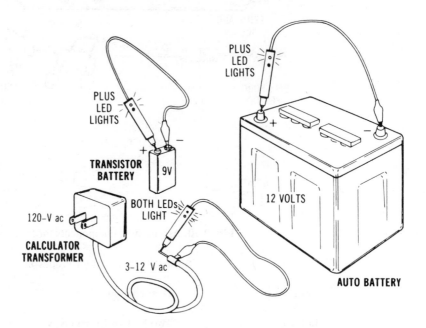

Fig. 2-11. Probevolt connected across voltage source.

Using the Probevolt

You will find the Probevolt a very handy test device for checking for the presence of voltages from about 1 to about 50 volts (ac or dc). Always connect the probe *across* a voltage source such as a battery, cell, or voltage power supply as shown in Fig. 2-11. It doesn't make any difference which lead you connect first to which source terminal since the probe will indicate and tell you which terminal is what. When the tip of the probe is connected to a positive voltage and the ground lead to the negative terminal, the plus LED will light while the negative LED will not light. Conversely, with the tip touching a negative terminal and the ground lead to a positive terminal, the negative LED lights and the plus LED does not.

Voltage Sensing

After you have used your Probevolt a short while you will notice that you can actually tell the condition of a battery by the brilliance of the LEDs.

Sense Battery Condition

For a given type of battery the higher the voltage, the brighter the plus or minus LED will be, and the better the condition of the battery. Let's take a look at some of the standard type batteries and see how we can use the probe to tell us something about their condition.

1.5-Volt Cell

Depending on the type LED used in the probe you may or may not be able to see the faint glow of the LED when connected across a 1.5-volt cell. If a cell is fully charged, however, you should see a faint glow if the voltage is above about 1 volt.

9-Volt Transistor Battery

A fully charged 9-volt battery will cause either of the LEDs to glow brightly. The Probevolt is very convenient for checking these batteries because you can connect the device to either terminal and still get an indication without having to reverse probes (because of incorrect polarity), which you would do using a voltmeter. You will soon know which batteries have enough charge to use and which ones to charge or discard. If you have a battery charger that charges cells of the C and D type, it can also be used to recharge the 9-volt transistor type. When you suspect that a battery is low, recharge it for two to three hours and retest. If the LED is brighter, use the battery. If not, discard it as it is not chargeable. If you recharge a 9-volt battery before it gets too weak, you will be able to recharge it two or three times before throwing it away.

12-Volt Auto Battery

There is an interesting thing you can do with the Probevolt and a car battery installed in the car. An automobile battery is designed to put out so much current that it is difficult to tell when it is good. It doesn't take much current to light an LED (leading you to believe the battery is okay), but when you try to start the car the battery is dead.

You can put the probe to very good use in looking for unknown discharge paths on the battery itself. An automobile battery accumulates a lot of road grime, oil, grease, dirt, and other matter, so that in a short time it has a low-current discharge path from the plastic casing to the automobile chassis ground. Clip the negative lead of the Probevolt to the negative lead of the battery or to any good, well-grounded screw (Fig. 2-12). Start at the negative lead of the battery and slide the tip of the probe along the top side of the battery casing. As you move away from the negative terminal of the battery (0 volts) toward the positive terminal you will see the positive LED begin to glow. Halfway toward the positive terminal the LED will be half lit (6 volts), and as you get to the positive terminal area the LED will be brightly lit (12 volts). You will be surprised to see the LED glowing when it is rubbed along the plastic housing! These are trickle discharge paths and point up the importance of keeping your battery clean and free of grime.

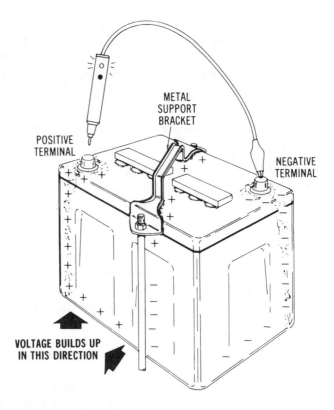

Fig. 2-12. Trickle discharge paths on a 12-volt car battery due to grease and road grime.

This slow trickle discharge is always overcome by the rapid charge of the alternator, but it will shorten the life of your battery. The situation can also be aggravated by a discharge path set up between the battery positive terminal and the metal battery support. Corrosion can occur on this metal support even though it is covered with a plastic coating. This condition will also cause the battery to trickle discharge, leading to a premature demise of your car battery.

Polarity Sensing

Let's take a look at how we can sense the polarity of a power supply. A transformer is used in most pocket-calculator battery chargers to change the 120-volt ac of the house current to a nominal 8 volts ac which is used to charge the nickel-cadmium batteries in the calculator. When you connect the Probevolt across the connector which plugs into the calculator, both LEDs will light if the voltage is ac. This is shown in Fig. 2-13. In a pocket calculator in which the battery pack is supplied with a nominal 6 to 8 volts ac the diode rectifiers are located in the calculator along with the battery pack. If, however, the diodes are located in the charger, 6 to 8 volts dc will be supplied to the calculator. You will be able to observe this by using the Probevolt. That is, you will be able to tell what is inside the charger by observing which LED lights.

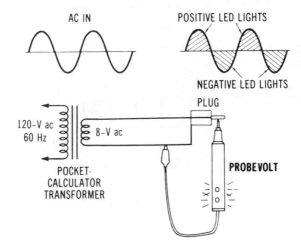

Fig. 2-13. Both LEDs are lit on an ac voltage.

In order to tell polarity of voltage with the probe, connect the clip lead (negative) to ground and probe with the tip. If the positive LED lights, you are across a positive voltage. If the negative LED lights, the probe tip is connected to a negative voltage. You can thus sense polarity without changing probes the way you have to when you test with a regular voltmeter. If both LEDs light, the voltage is ac, sine wave, or square wave.

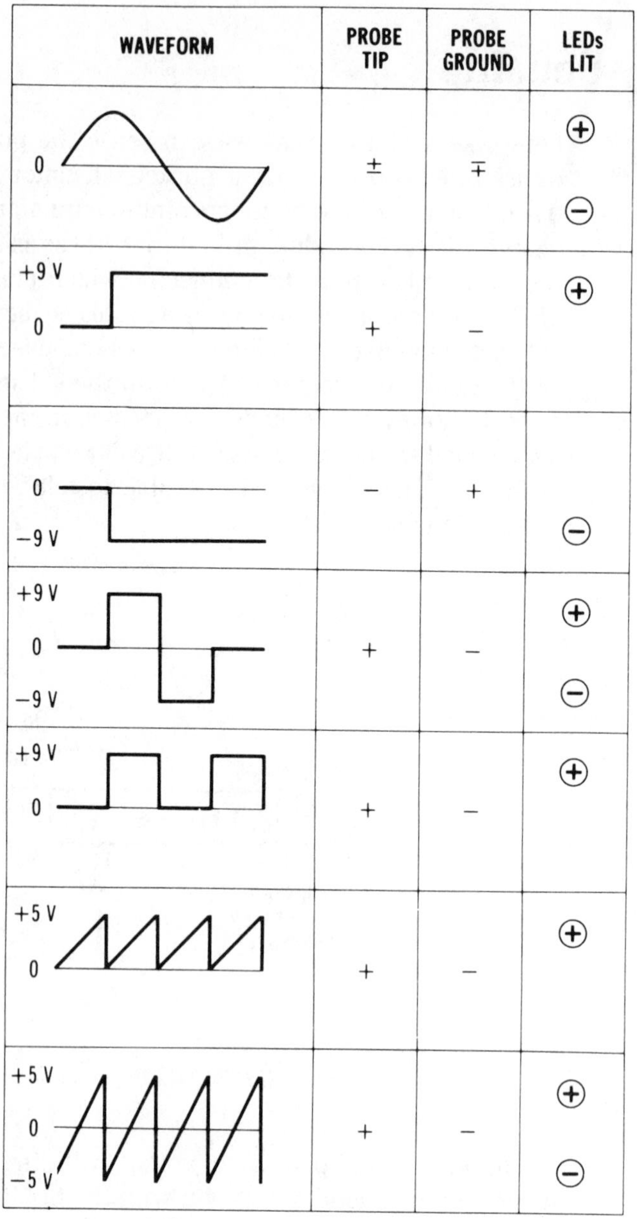

WAVEFORM	PROBE TIP	PROBE GROUND	LEDs LIT
0 ⟋⟍⟋ (sine wave)	±	∓	⊕ ⊖
+9 V, 0 (positive step)	+	−	⊕
0, −9 V (negative step)	−	+	⊖
+9 V, 0, −9 V (bipolar square wave)	+	−	⊕ ⊖
+9 V, 0 (positive square wave)	+	−	⊕
+5 V, 0 (positive sawtooth)	+	−	⊕
+5 V, 0, −5 V (bipolar sawtooth)	+	−	⊕ ⊖

Fig. 2-14. Probevolt LED response to various waveforms.

Response to Different Waveforms

Some representative waveforms, along with the LEDs that would light to indicate their presence, are shown in Fig. 2-14. In the event both LEDs are on but one LED is much brighter than the other, it indicates the waveform is not a sine wave but that it has a greater voltage swing in one direction than the other. The duty cycle of the waveshape also determines the LED brightness. The greater the duty cycle, the brighter is the LED for that polarity of voltage.

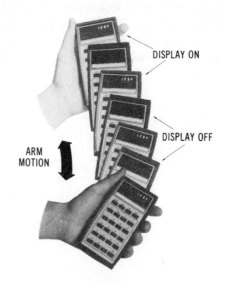

Fig. 2-15. Calculator display flashes on and off.

The Moving LED

When we look at the LEDs of the Probevolt, the display of a pocket calculator, or a quartz digital watch, the LEDs appear to be on continuously. It is easy, however, to show that they are not and that they are being multiplexed, or switched on and off, at a rate so fast that to the eye they appear to be on all the time. It is possible to observe visually that the LED display is being multiplexed by holding a pocket calculator firmly in your hand while making a rapid back-and-forth motion with the arm as it is swung in a short arc. Fig. 2-15 shows the number 123.4 being displayed while this movement is being made. You can darken the room slightly in order to show this to full effect. The row of digits being displayed will appear to jump in a "Keystone Cops" strobelight manner as the arm is swung rapidly back and forth. The on and off times of the digits can be observed by comparing height or size of the digits to the distance traveled during the off times. Most displays are on for 50 percent of the time. We will discuss the LED in motion in greater detail in a later chapter.

Circuit Loading

The LEDs will draw 5 to 25 milliamperes of current when in operation and may therefore load down any circuit to which the probe is connected. The Probevolt will not affect most circuits to which it is connected, though it may load down some digital integrated circuits and stop them from operating properly. You will quickly learn which type circuit it will not affect adversely.

3 LED Voltage and Polarity Indicator

This voltage test device is similar in operation to the Probevolt described in Chapter 2 but is designed so that it operates over a greater voltage range of 2 to 120 volts ac or dc.

Operation

The voltage level tester is as simple to operate as the Probevolt is. However, it uses two test leads with alligator clips so the unit can be connected in a circuit and left there for a period. Like the Probevolt the level tester is connected in parallel, or across, the voltage to be observed. Fig. 3-1 is a photograph of the voltage level tester mounted in a plastic case which is available at local electronics stores.

Circuit Analysis

The circuit for the voltage level tester is shown in Fig. 3-2. A switch is used to increase the total resistance in the circuit in steps so that increased voltage is dropped across the resistors and the voltage drop across the LED remains more constant. We can compute the voltage drop necessary across each resistor by using Ohm's law, a nominal LED current of 20 mA, and an LED voltage drop of 1.6 V. For R1 we have:

$$R1 = \frac{2 - 1.6\,V}{20\,mA}$$

$$= \frac{0.4\,V}{2 \times 10^{-2}\,A}$$

$$= 20\,\Omega$$

In like manner we can compute the values of the remaining resistors which are given in Table 3-1. However, since inexpensive resistors of the exact values as shown are not readily available, a column is given for standard 20-percent resistors. These values will do the job adequately and are available at local electronics stores. Wattage sizes for the resistors are computed according to the expression:

$$P = I^2R$$

where
   P is the power dissipated in the resistor in watts,
   I is the current through the resistor and LED in amperes,
   R is the value of the resistor in ohms.

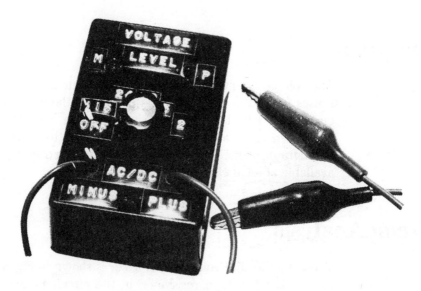

Fig. 3-1. Voltage level tester mounted in plastic case covers wide voltage range from 2 to 120 volts ac/dc.

Note that all resistors are ½ watt except R5, which must be 2 watts for continuous operation at the 120-volt switch position.

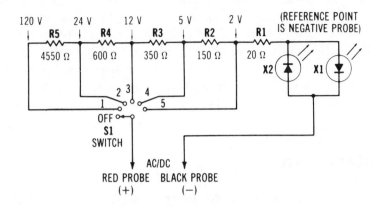

Fig. 3-2. Voltage level tester covers range of 2 to 120 volts through resistors that are switched into circuit.

Table 3-1. Resistor Values and Power Ratings

Resistor	Value (Ohms)		Power (Watts)	
	Exact	Use	Exact	Use
R1	20	22	0.008	½ W
R2	150	150	0.06	½ W
R3	350	330	0.14	½ W
R4	600	560	0.24	½ W
R5	4550	4700	1.80	2 W

Wide Voltage Range Operation

The Probevolt covered one range of voltage and the LEDs became brighter as the voltage in the circuit became greater. This was due to the greater amount of current in the circuit. In the voltage level tester we cover a wider range of voltages using different values of resistors, starting at 120 volts first (most resistance in the circuit) and ending at 2 volts (least resistance in the circuit). Most electronic devices operate at 120 volts ac or less, so we can safely use the tester on most solid-state, transistor, and IC circuits.

Before you connect the voltage level tester into a circuit, switch it to the off position. Then carefully connect the tester into the circuit to be tested and switch from 120 volts to a lower voltage range until the LEDs start to glow at a nominal brilliance, which you will soon get to know. For an automobile, we should switch to the 12-volt range (12 volts dc). For a central air conditioner or heating unit, we should switch to the 24-volt range (24 volts ac) when testing the control circuit. A good 9-volt transistor battery is a convenient battery to check on the 12-volt range to determine relative brilliance so that you can tell if a 9-volt battery is still useful. So keep a good 9-volt battery nearby. Remember, the level tester gives a relative voltage indication of an ac or dc voltage. A voltmeter can give an exact value but it cannot easily give you an indication of the polarity and duty cycle of a voltage.

Construction

The parts required for construction of the level tester are shown in Table 3-2.

Table 3-2. Parts List for Voltage Level Tester

Item	Description
X1, X2	Red or Green LEDs, 2 each
R1	Resistor, 22-Ω, $^1/_2$-W
R2	Resistor, 150-Ω, $^1/_2$-W
R3	Resistor, 330-Ω, $^1/_2$-W
R4	Resistor, 560-Ω, $^1/_2$-W
R5	Resistor, 4700-Ω, $^1/_2$-W
S1	Switch, Rotary, 6-Position
Case	Plastic Case, Radio Shack 270-230 or Equivalent
Leads	Test Leads, Insulated with Clip, 36 inches (90 cm) long
Misc.	Knob, Solder, Wire, Labels

The Parts

The level switch can have any number of switch positions so long as it has six positions including off. The LEDs can be red or green so long as they are both the same color as it is important that their brightness is the same on both sides of an ac voltage sine wave. You can match several LEDs out of a small group of four or five of the same color. Remember, the red LED has a forward voltage of 1.6 volts dc, while a green LED has a nominal 2.0 volts forward voltage. You can connect a red and green LED in parallel (reverse polarity) but their brilliance will have less meaning when using the level tester on an ac voltage as they will light to different intensities. If, however, you match intensities of a red LED and a green LED and they look good to you on an ac signal, use them that way. If you use the green LED for X1 and red for X2, then a green light would indicate a positive voltage and a red light would indicate a negative voltage, and for this case with both LEDs on, an ac voltage is indicated.

Packaging

Fig. 3-3 shows the locations of the five holes to be drilled in the plastic case. Drill two small holes for the two LEDs and one larger hole for the six-position switch. Drill two small holes just large enough for the test leads to go through. Press the LEDs gently into the two holes and use an epoxy glue to hold them in place if necessary. You should not have to use any additional hookup wire to connect the circuit as the resistors and LEDs have leads long enough to reach the necessary points. When wiring make a mechanical connection first and then apply a small amount of solder to all points to ensure good electrical connection. Remember to try different placement of the wires before you make a final mechanical connection involving bending the leads as the actual layout of components is nothing like the schematic diagram! This is one of the interesting things to learn in electronic kit building—how to arrange the components with the least amount of confusion so that they can be connected together according to the simple-minded circuit schematic diagram.

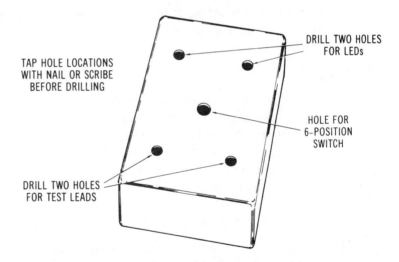

Fig. 3-3. Hole locations in plastic case for the voltage level tester.

Labeling

Press-on strip labeling from hand-operated machines is available at most radio stores, the local discount stores, and most supermarkets. These machines are ideal for making labels which you will put on the devices that you build. The labels help give the appearance of a job well done and they also identify the various devices and operating controls to friends who may use them. Black labels on black plastic boxes look really good but you may use other colors that match the box into which you finally put your device.

Using The LED Indicator

Before using the LED level indicator for checking voltage levels, you should check the wiring and test the indicator as described in the following.

Check Wiring

Make certain you have wired the circuit correctly according to the diagram. You'll be surprised how often you'll forget to complete part of the circuit, forget to solder a connection, or make an incorrect connection which causes the circuit to malfunction. It is worthwhile to use a yellow, pink, or light-blue marker or highlighter to indicate on a circuit diagram that portion of a circuit which you have completed. This will save you from making one connection twice, while forgetting to make another connection once! Remember, it is not because you are forgetful but it is due to the complexity of some electronic circuits.

Testing the Unit

After you have checked the wiring, put the aluminum cover on the plastic box so there are no exposed wires or leads. You might come across 120-volts ac, so make the level tester connections into the circuit with the voltage to be tested turned off. It's best not to clip into a live 120-volt circuit as it is too easy to get shocked, or you might drop a lead and blow a fuse or trip a circuit breaker. There is no bodily harm from the other lower-voltage levels, but you need to exercise caution when you tap or clip into a circuit, as you might short out some component or voltage source.

Again, the 9-volt battery is a convenient reference to have around the work bench as it will provide 5 to 9 volts for several years. Place the alligator clips across the battery terminals and then adjust the voltage level switch; one of the LEDs should get brighter as you switch closer to 12 volts. If you have connected the positive probe to the smaller (plus) male terminal of the 9-volt battery and the negative probe to the larger (female) terminal, the plus or positive-marked LED should be lighted. The LED marked negative will be out completely. When you switch to the 5-volt or 2-volt range, the LED will become very bright so you should quickly switch back to the 12-volt range. Now you have checked out all the voltage level ranges, and you are now ready to use the device.

Voltage Sensing

The level tester is very simple to operate; it tells you easier than a voltmeter does, and you don't have to reverse test leads as you can connect either test lead to a voltage source and still sense voltage level as well as something about its polarity. It is very handy for checking power supplies of various voltage levels as well as the polarity of a source. Fig. 3-4 shows an IC chip powered by a plus and minus 9-volt supply and how the level tester quickly tells you all you want to know about the battery supply. By placing the negative probe at point *A* (Fig. 3-4) and the positive probe at point *B* the positive LED will light, giving a relative showing of the positive 9-volt battery condition. When you move the positive probe to point *C*, the negative LED will light, showing that the probe tip is connected to a negative source (relative to the negative probe), and the negative LED will light to the brightness of the negative 9-volt battery. You can then move the negative clip to point *B*, and the negative LED should be considerably brighter as now the total applied voltage will be 18 volts, depending on the condition of both of the batteries. You will soon get to know the meaning of the LED relative brightness.

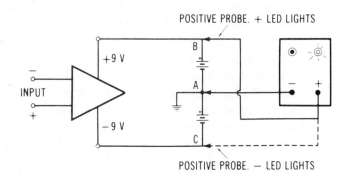

Fig. 3-4. Voltage level tester connected across 18-volt battery supply.

Polarity Sensing

As was discussed in the chapter on the Probevolt, you will be able to tell what the polarity of a voltage terminal is by noting which LED lights. If both LEDs light, they tell you that the level tester is connected across an ac voltage, such as a calculator power supply transformer (6- to 10-V ac), a 6-volt filament transformer (6.3-V ac), a doorbell transformer (18-V ac), a central-air heating or cooling power supply system (24-V ac), or a model-train power transformer (0- to 10-V ac).

Operation on AC, DC, and Audio Voltages

The unit will operate on the range of 2 to 120 volts, ac or dc, and it may be used to display audio voltages.

AC Voltages

We know that when we connect across an ac voltage supply that both LEDs should light, because part of the time the voltage is positive and part of the time it is negative. We will, however, now look into a case of something that is "not all that meets the eye." Most model-train power supplies (usually a transformer with a variable tap for speed control) also contain a diode rectifier so that the unit can supply direct current to the train track. So, when we connect the level tester across the transformer output which goes to the tracks, we see that just one of the LEDs lights. Which one depends on the probe connections we have made to the track. We haven't learned much so far, except that we have dc voltage out of the transformer. But with a little trick we will turn the inexpensive voltage level tester into a $200 oscilloscope! That is, it would take an oscilloscope to observe the voltage waveform we will be able to see with the tester. With the probes still connected to the transformer, gently move the tester back and forth in a 6-inch to 12-inch (15- to 30-cm) arc. You will notice that the one LED appears to flash on and off. With the tester stationary the one LED is on all the time, showing that we have a dc voltage applied. But that is the power of the moving LED; it can give you a dimension other than just that of voltage—namely, the dimension of time!

The Moving LED

Fig. 3-5 shows the circuit for a simple train or calculator power supply. Note that we are using half-wave rectification and that the ac waveform is converted into a *pulsating* dc waveform. The stationary LED appears to be on all the time as shown at *a*. But with the LED moving as in *b* of Fig. 3-5, the eye sees that the LED is turned on and off. That is, the LED moves through space while it is off, not creating any trail of light. When it turns on again it creates such a trail. So we see the "Keystone Cops" flicker effect and, more recently, the flashing effects of the giant strobelights used in modern musical rock-group stage presentations.

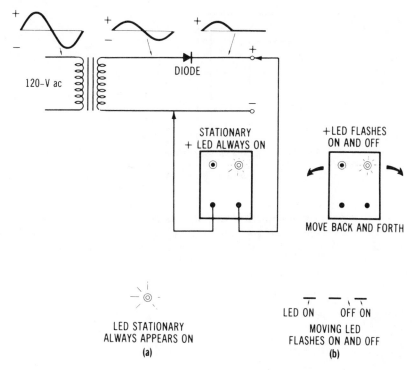

Fig. 3-5. With an ac voltage applied to the primary, the LED appears on (a) when tester is stationary and it flashes on and off (b) when tester is moved.

DC Voltages

When you connect the voltage tester or Probevolt to a dc source, such as a 9-volt battery (Fig. 3-6A), and move it back and forth, the LED will remain the same intensity the whole movement. That is, it won't flash on and off. Fig. 3-6B shows the trail of an LED movement with dc voltage applied to the tester. This will tell you for sure that you have a dc voltage.

Audio Voltages

A high power stereo amplifier will make the voltage level tester come alive! When you connect it across the speaker terminals it will really flash. The 5- or 2-volt range is best for most room-speaker volumes, and you will be able to "see" the ac voltages which music or voice sounds make. You will often see that, even with the LEDs stationary, the sounds produced are not symmetrical. That is, the positive or negative LED will flash more than the other. And when you move the tester with music playing, you will create a

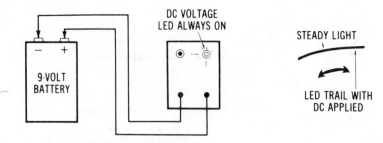

(A) Voltage tester connected to 9-volt battery.   (B) Trail of LED when tester is swung.

Fig. 3-6. Indication when tester is connected to a dc voltage.

most interesting display of musical light. You will soon want to add yellow and green LEDs to the red device you now have so that you will have a flashing multicolored spectacle of sound!

4 A Solid-State Telephone Bell: The Sonabell

It was only in recent years that the sound of a ringing telephone has changed. The ring of a phone bell is very distinctive, but there are times when you would like it not to be so harsh, or so loud, or maybe even not on at all. The Sonabell gives you these choices. Its sound is rather musical—like that of the chirp of a bird. Fig. 4-1 shows the Sonabell mounted in a plastic box convenient enough to move any place. A volume control permits you to adjust its volume from full on to off. In a small room you don't need much volume anyway. "Sonabell" is derived from *sonus* (sound) and *bell* (bell).

Operation

The Sonabell is plugged into a telephone outlet at any location in the house, and it will sound out whenever the phone rings, no matter where the bell ringer is. You can even turn the regular bell all the way down and use only the Sonabell. It will not affect normal operation of the phone or the phone line. The Sonabell is very handy to connect to a phone outlet in the garage or on the patio as you can run it a great distance without affecting its proper operation. With the Sonabell outdoors or in an open garage you will be able to hear its melodious call hundreds of feet away. If you are swift of foot you will be able to answer the phone before it has rung four times. And because of birdlike chirping tones of the ring produced by the Sonabell, you will not disturb your neighbors; your "ring" will sound much like all the other birds chirping, except you will recognize its tone and its repetition. If a Klaxon™

Fig. 4-1. Sonabell mounted in plastic case with extension plug.

horn were used to call you, you would soon lose all your friends in the neighborhood.

Circuit Analysis

A look at the circuit diagram in Fig. 4-2 shows the Sonabell connected in parallel across a telephone line. It is in parallel with the "main" phone in the same manner that an extension phone would be. The 48 volts dc across the 47-kilohm resistor and neon bulb does not cause any action as the neon bulb is not ionized. The neon bulb thus appears as an open circuit at this time. When the phone rings, however, a voltage of 90 volts rms at 20 Hz (actually 20 pulses per second) appears across the line and causes the regular electromechanical bell to ring. That is, a clapper moves back and forth, striking a bell on each end of its stroke, giving the familiar phone-ring sound. At this time part of the 90-volts rms appears across the 47-kilohm voltage-dropping resistor and part across the neon bulb. Most neon bulbs ionize with 65 volts across their electrodes and, when one is so ionized, current will flow through the bulb, through the resistor, and the ionized neon gas emits light that we see as an orange glow. The emitted light is received by R3 photoresistor (light-sensitive resistor) which rapidly changes its resistance value from a nominal 10 megohms to a nominal 1200 ohms.

Looking now at the sound-producing part of the Sonabell we see a series circuit consisting of the piezoelectric sounder or annunciator (there is no word coined for the piezoelectric sounder device—it is not a buzzer, not a bell, not a speaker, so we'll call it the piezosounder), a 9-volt battery, the photoresistor, the LED, and a volume control. When the 50-kilohm volume control is set for full output from the piezosounder and the phone bell rings, the photoresistor drops to a nominal 1200 ohms, and the sounder puts out at full volume. It chirps each time the phone rings. You will find the output to be very pleasant and truly a solid-state sound.

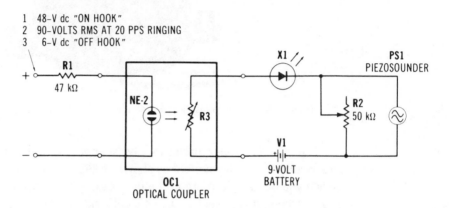

Fig. 4-2. Circuit diagram for Sonabell.

Piezoelectric Sounder

You may have a piezoelectric-effect device on your wrist and not even know it. If you have a watch controlled by a quartz crystal you have a very stable timepiece because of the special property of the crystal. You also have a quartz crystal in your CB transceiver, your batteryless electric cigarette lighter, and in the automatic clock-controlled thermostat for heating and cooling your home. The crystal has a special property: that of piezoelectricity. If a piece of quartz crystal is placed between two electrodes it is mechanically strained when a voltage source is connected across the electrodes. Conversely, when the crystal is squeezed between the two electrodes it will produce a voltage across the electrodes. That is, the piezoelectric crystal can be used to transform mechanical energy into electrical energy and electrical energy into mechanical energy.

The Mallory Type SC628 is a solid-state audible warning device which makes use of the property of the crystal when we apply a voltage to the electrodes and mechanical energy (sound) is produced. The SC628 emits the sound you usually hear at airport passenger security checkpoints and from many other electronic devices. It emits a very pure sound at 2900 Hz when between 1 and 39 volts dc are applied to its terminals.

The Radio Shack piezosounder (Cat. No. 273-060) is shown in Fig. 4-3 and it emits a pure signal at 3600 Hz. It is a fairly inexpensive unit and operates at a very low current from less than 4-volts dc to about 28-volts dc, the sound output increasing with voltage applied.

Fig. 4-3. Solid-state piezoelectric sounder. *Courtesy Radio Shack, Div. of Tandy Corp.*

There is no proper name for the sounder, so we have coined the term *piezosounder* and the symbol as shown in Fig. 4-2 for its circuit diagram. This device will be used in more and more devices of the future for home, at work, and at school. It is novel because it is all solid-state, there are no moving parts, no arcing electrical points as in a buzzer to cause an explosion in gaseous environments or to cause radio-frequency interference, it consumes very little electrical power, and it will operate on just a few milliamperes. This very special device will be used in other circuits in later chapters.

Optoelectronic Connection

The Sonabell must be connected to the telephone line at some point in order to tell when the line is ringing. We do this through a technique which permits complete electrical isolation between the input (telephone line) and the output (the piezosounder driving circuit).

The Lady Bug

A device that provides isolation as we need is called a *photocoupler* or *optical coupler*. Fig. 4-4 shows the schematic for three types of light emitters. These optical couplers are available from a number of suppliers, with the Lady Bug being available from Sigma Instruments, Inc., 170 Pearl Street, Braintree, MA 02185. Fig. 4-4A shows an incandescent lamp as the light source and a photoresistor which rapidly and smoothly reduces the resistance of the cell when the light strikes its surface (usually cadmium sulfide: CdS). The lamp is slow to turn on as it must warm up to give off light. Fig. 4-4B shows an LED

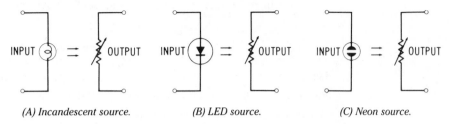

(A) Incandescent source.   *(B) LED source.*   *(C) Neon source.*

Fig. 4-4. Optical couplers using photoresistor sensors.

light source, which has a fast response time but operates at a low voltage. In Fig. 4-4C we see a neon bulb as a light source which has a rapid turn-on.

Fig. 4-5 shows the rapid turn-on time for a neon light source with its photoresistor. A photoresistor has a rapid turn-on rate but its turn-off rate is slower, as we see in Fig. 4-5. Some photoresistors have a "dark" resistance of 10 megohms but it may take 10 seconds to reach this value. Some characteristics of the Sigma Lady Bug we will use are shown in Chart 4-1. It is to be noted that these type devices can be identified by a number of terms, such as optical coupler, optoelectronic relay, and optical isolator. It is *essential* that the 47-kilohm series dropping resistor be used as it provides protection for the neon bulb when it ionizes and conducts. If it is not used, the neon bulb will be destroyed due to the high current through the ionized neon gas. In fact, neon can conduct so much current the copper leads may melt before the neon is destroyed.

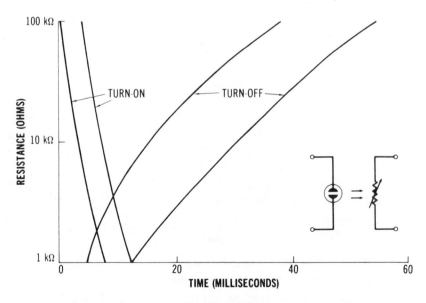

Fig. 4-5. Typical turn-on and turn-off for an optical coupler
using neon bulb and photoresistor.

Chart 4-1. Characteristics of Sigma 301T1-120A1 Optical Coupler

Lamp	Photocell
Ionization voltage: 78-V ac rms	Max. On Resistance: 1200 Ω
Rated life: 15,000 hrs @ 1.3 mA Average	Min. Off Resistance: 10 MΩ
Isolation—Lamp to Cell: 1000-V ac rms	Max. Turn-On Time to 10 kΩ: 15 ms
	Max. Turn-Off Time to 100 kΩ: 75 ms

Make Your Own Coupler

You can make your own optical coupler by "connecting" a small neon bulb, such as an NE-2, and a photocell, both available from your local radio electronics store. Fig. 4-6 shows how the neon bulb and photocell are mounted in a small section of a cardboard or plastic tube. Place them close to each other till they touch, and fix them in place with epoxy or glue. Seal the tube ends so that no light will be able to enter. Wrap the ends tightly with black electrical tape. You can identify the two neon-bulb leads with an ohmmeter or continuity tester because the bulb will have infinite resistance when it is not ionized. The leads from the photocell will indicate a resistance of between 1 and 10 megohms in the dark condition (no light on cell).

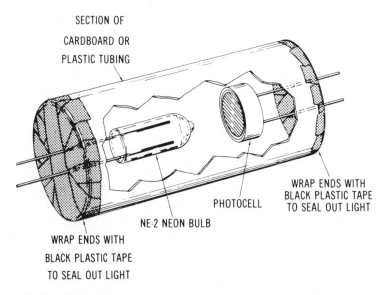

SECTION OF
CARDBOARD OR
PLASTIC TUBING

WRAP ENDS WITH
BLACK PLASTIC TAPE
TO SEAL OUT LIGHT

PHOTOCELL

NE-2 NEON BULB

WRAP ENDS WITH
BLACK PLASTIC TAPE
TO SEAL OUT LIGHT

Fig. 4-6. Neon bulb and photoresistor close to each other to form optical coupler.

A Ringing LED

An LED is connected in series with the piezosounder and it will flash at a 20 pulse-per-second rate when the phone rings. The circuit is arranged so that as you turn the volume control all the way off, the LED will be at its brightest. You can easily see the LED flash, especially if you have the Sonabell sitting on your desk. The LED feature is most helpful when you want to turn the phone ringing and the Sonabell all the way down. You'll have silence but not miss any calls coming in. The battery in the Sonabell will last for several years (its shelf life) because there is very little power consumed during ringing and none during quiet conditions.

Construction

The parts of the Sonabell are mounted in a plastic case with the parts arranged in a manner suggested by Fig. 4-1. You may have a certain color box or case that will better suit your needs and, if you want maximum volume all the time, the volume control can be omitted. Table 4-1 lists the parts required for building the Sonabell; all these parts can be found at the various electronic parts stores.

Table 4-1. Parts List for Sonabell

Item	Description
V1	9-V Transistor Radio Battery
X1	Red or Green LED
OC1	Optical Coupler, Sigma 301T1-120A1 (If not used, see NE2 and R3.)
NE-2	Neon Bulb, NE-2 (Use if OC1 is not used)
R1	Resistor, 47-kΩ, $\frac{1}{2}$-W
R2	Volume Control, 50-kΩ
R3	Photocell, Cadmium Sulfide (CdS) (Use if OC1 is not used)
PS1	Piezosounder, Mallory SC628 or Radio Shack 273-060 or Equivalent
Misc.	Telephone Plug for Parallel Connection, Plastic Case, Knob for Volume Control, No. 22 Wire to Reach Phone Jack (Pair), Clip for Connection to Battery.

Use a circle cutter to cut a round hole in the plastic case for mounting the piezosounder. If other than a Sonalert is used, fix it in place with epoxy after drilling a hole about $\frac{1}{2}$ inch (1.27 cm) in diameter for the sound to radiate from the unit. Drill a small hole for the LED to protrude and fix it in place with epoxy or glue. The volume control is mounted in the center of the case below the sounder. Wire up all connections and double-check them to be sure you've made them all. Remember, use a colored marker to indicate on the schematic that portion of the circuit that has already been wired. Remember also that you must observe the marked polarity of the piezosounder. The black wire is ground (minus) and red wire is positive. Observe the polarity of the LED and always connect the anode (+) to the positive side of the battery circuit. The cathode (−) of the LED is connected to the negative side of the circuit.

Connection to Telephone Line

Most homes and offices are equipped with jacks so phone instruments can be moved from room to room. In Fig. 4-1 an older four-prong jack is shown. These jacks will also accommodate a parallel jack so that two instruments can be plugged into the same outlet. Many systems today use the newer "modular" jack and plug system. Dual jacks are readily available to allow two plugs to be used. Be sure to use the type of plug which matches your phone system. It is at this location that you connect the Sonabell most easily into the circuit. Most single-party lines make use of the red and green pair of wires with 48-volts dc appearing across the pair when measured with a voltmeter. Reverse the connections from the Sonabell to the jack and plug back into the line if the bell does not appear to function properly when the regular bell rings. The yellow and black pair of wires in the four-conductor telephone cable in your house are used for other purposes and do not carry any ringing voltage. Therefore do not connect into them.

Using the Sonabell

To test the working of your Sonabell, call a friend and have him or her call you back. Let the phone ring enough so you can observe operation of the LED (it should flicker each time the bell rings). The volume control should be connected so that there is minimum volume with the control rotated counterclockwise, and fully on with it clockwise. When volume is on full, the LED will not be so bright since most of the battery voltage will be across the piezosounder. Of course, there is always some of the battery voltage across the photocell in the ringing condition.

You soon will become used to the pleasant musical sound of the Sonabell and probably will prefer it to the regular phone bell. If you have a phone outlet on the patio, garage, or attic, but not an instrument, you can plug the Sonabell into the outlet and work or study in comfort knowing you will be able to hear the phone ring. Even with a very large yard you will be able to hear the Sonabell on the patio or back porch and come running and answer the phone before you have heard three or four rings. Yet the Sonabell will not disturb the neighbors, because of its musical sound, like that of a bird chirping.

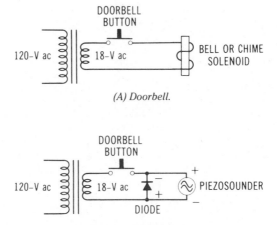

(A) Doorbell.

(B) Piezosounder circuit.

Fig. 4-7. Doorbell replaced by piezosounder.

Additional Uses for the Sonabell

The Sonabell can also be used to replace your present doorbell or chimes if you would like a change in the sound of the doorbell. Fig. 4-7 shows how the piezosounder is connected across the two wires that go to your present bell-activating circuit. Simply remove the two wires to the bell and mount your new sounder in its place or alongside it. Then connect the bell wires to the piezosounder. Additionally you can wire your new bell in parallel with the old bell and place it in the garage, attic, den, or even on the patio to let you know that someone is ringing your doorbell. Note in Fig. 4-7 that a diode is used to protect the piezosounder since the 18-volt bell voltage is ac. The diode conducts when the waveform is negative. The piezosounder will chirp at a rate of 60 pulses per second when the doorbell button is pushed.

5 The Poweralert: A Line Voltage Monitor

A power failure may not happen very often in your part of the country or community, but when it does and you don't know about it you can be mighty upset, frustrated, late, or all of the above.

Operation During Power Failure

The Poweralert line voltage monitor operates off the ac power lines, and, when they fail, the Poweralert will awaken you so that you can take some alternative action. Fig. 5-1 shows the Poweralert mounted in a plastic case which can be placed near the bed within reach for alarm shutoff.

In some parts of the U.S., brownouts may occur during periods of extreme heat or cold. There are also power interruptions. The interruptions may be unintentional (storms, severe weather, lightning strikes, etc.) or they may be intentional on the part of the power company in order to protect the power supply system (conservation, equipment repair, fuel conversions, trading of power, etc.).

How a power failure affects you depends a lot on what time of the day or night it is and what you are doing at the time. The line monitor is most useful when you are asleep. If you use an electric alarm clock to awaken you in the morning, a 5- or 10-minute power failure may not affect you much as you will only be 5 to 10 minutes behind your regular schedule, and this you can easily make up. If you use an electronic digital clock and there is a power failure, however, when the power comes back on, the clock does not resume normal timekeeping. It merely flashes its numerals, colon, or the like, and you remain fast asleep. If you have no-break power (nickel-cadmium battery backup), timekeeping will continue during the power failure and the alarm will sound on time.

A regular alarm clock of the electric type will continue timekeeping when the power comes back on. If the power has been off two to three hours, however, and there is nothing to awaken you, you will obviously be late in getting up unless you have a mechanical, spring, or weight-driven clock and listen to the sounding of the hour. Therefore, the Poweralert is often a useful device.

Fig. 5-1. The Poweralert will waken you during a power failure.

Circuit Analysis

The circuit diagram for the Poweralert is shown in Fig. 5-2. The main item in the circuit is a 125-V ac relay, K1, which is held in the *on* condition by the presence of 115-V ac at any convenient wall outlet. X2, a green LED, is in series with the relay winding of K1. The resistance of the winding (about 4500 ohms) serves as a current-limiting resistor for X2. Diode X1 protects X2 from reverse polarity voltage, which appears across X2 during the negative swing of the ac voltage. The green LED lets you know it is plugged in and also helps you find the Poweralert in darkness. It also serves as a nightlight.

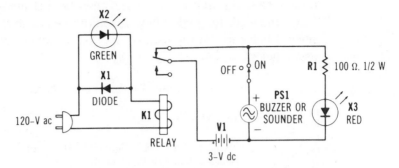

Fig. 5-2. Circuit diagram for voltage monitor and alarm.

During a power interruption or failure that lasts about 0.2 second or longer, the relay will de-energize. This places the standby part of the alarm into action and the buzzer or piezosounder will be switched on as it is energized by the two size AA batteries. The AA cells are capable of supplying power to the buzzer and red LED for a number of hours, while their shelf life is about a year or so.

When the power fails, the buzzer will sound and you will want to shut it off. The red LED (X3) helps you locate the unit if you have placed it on the floor under your bed. Remember, it will be pitch black at this time because all other lights will also be out except for the red LED unless you have a no-break battery light system. You can then shut off the buzzer PS1 by turning switch S1 off and the red LED will remain on. Resistor R1 serves as a current-limiting resistor for the red LED. The Poweralert is left permanently plugged in and consumes about 3 watts of power, the same as an electric clock.

Construction

The line voltage monitor can be built into any handy case and the components mount neatly in the small plastic cases available locally. Table 5-1 lists the parts required to build the Poweralert. Use a low-current 125-V ac relay with single-pole, double-throw contacts, and, remember, it is the relay off contact that sounds the alarm and lights the red LED. Arrange the components as shown in Fig. 5-1 and drill small holes for the LEDs. Push them gently through the plastic case and hold them in place with a small amount of epoxy or glue. If you use the slide switch specified for turning the buzzer on and off, drill three or four small holes in a row and then file them out so the slide switch can be mounted and still have enough movement for proper operation. Drill a hole through the plastic case for the line cord and pass the tinned leads through the case and then tie a strain-relief knot in the cord. Leave the leads long enough to reach the relay and the LED terminals. Drill several holes in the top side of the box so sound from the buzzer can be easily heard.

Table 5-1. Parts List for Poweralert

Item	Description
V1	Two Size AA Cells in Battery Holder With Clip Connections, Radio Shack 270-382 or Equivalent
PS1	1½- to 3-V Buzzer or Piezosounder
X1	Diode, 1N4001, 1N914, or Equivalent
X2, X3	Green and Red LEDs
R1	Resistor, 100-Ω, ½-W
K1	Relay, 125-V ac, Radio Shack 275-217 or Equivalent SPST Switch, Radio Shack 275-406 or Equivalent
Misc.	AC Line Cord with Plug and Utility Case, Radio Shack 270-231 or Equivalent, Solder, Labels

Solder all connections as you make them and mark them off on the circuit diagram so you will know that you have completed them. Then be sure to recheck all the connections before plugging the unit in.

Testing Operation

Insert two fresh AA cells in the battery holder and the red LED should come on and the buzzer sound with the unit not plugged in the wall. Operate the buzzer on/off switch to see that it is operating properly. Plug the Poweralert into an ac outlet and you should hear the relay click on and see the green LED come on and the red LED go out. Plug the unit in several times to see that the relay will indeed drop out and the buzzer sound when unplugged. After that it is a case of placing the unit in a location close to your bed where you can easily reach it during the night.

Wake up During Power Failure

During severe weather when there probably will be short power interruptions of 1 to 2 seconds the buzzer will come on and then go off. This will happen especially when there are nearby lightning strikes which short out the power line momentarily. As mentioned earlier, during a power failure at night the bedroom will be pitch black and you will be able to locate the Poweralert by the glowing red LED. The buzzer can be switched off, leaving the red LED glowing. If it is not time to get up, you may want to set a manual alarm clock to awaken you later even though the power does come back on shortly. If you are in a car pool, you may want to call others when you wake up in order that they also won't be late in the event the power outage was widespread and also hit their section of town. If you are catching an early airplane flight, you certainly don't want to miss it due to a power failure.

Check the unit every few months to see if the batteries are still strong enough to sound the buzzer. If you hear the relay click when you unplug the unit from the wall but don't hear the buzzer, the batteries are too weak and should be replaced.

Additional Uses for Poweralert

There are several other areas where the Poweralert is very useful.

Locate Circuit Breakers

The Poweralert can aid you in locating a circuit breaker or fuse which controls a specific outlet or lamp you wish to disconnect in order to work on it. Plug the Poweralert into the outlet so that the buzzer cuts off. Then go to the circuit-breaker box or fusebox and one by one shut off the breakers (or unscrew the fuses), until you hear the buzzer sound off. You then know that you have located and turned off the breaker that supplies that specific circuit. You could plug in a lamp to do the same thing except you cannot see the lamp around a corner. But you can hear the buzzer through many rooms, down in the basement, or up in the attic.

Freezer Alarm

When the power to your freezer is accidentally cut off you could be in for a lot of trouble if you are not aware of the problem. You can plug the Poweralert in the same receptacle as the freezer, and if the power is ever accidentally cut off because the circuit breaker trips or fuse blows, the buzzer will sound and you'll hear it from a distance. Remember, you may not notice for days that the external freezer panel light has gone off. If your frozen food and meat defrost, you could lose hundreds of dollars worth of food. And what is worse, when food is completely spoiled because it has been defrosted for several days it could stink up your freezer so badly that you might never get the odor out of it. That could also mean a considerable loss.

Intruder Alarm

During the night an intruder may pull the main power switch to your house so that he or she can disconnect power being supplied to lights or various security guards which do not have a no-break power supply. In this case the Poweralert alarm would go off and awaken you. A glance out the bedroom window to see if any other lights are visible will help establish whether the power outage is confined to your house only or covers a neighborhood area. At any rate you are forewarned. It's a good idea to put your own lock on your main power cutoff switchbox so that you need a key in order to turn it off. You will still have full house protection from the line breakers or fuses.

6 A Light-Sensitive Audio Oscillator: The Sonalight

Most electronic devices change one kind of energy into another kind of energy. The common light bulb changes an electrical voltage (which we cannot see) into electromagnetic energy, light, that we can see. The speaker in a high-fidelity radio receiver system changes alternating voltage in the audio-frequency range (which we cannot hear or see) into acoustic sound that we can hear but cannot see. The *Sonalight*, which we will build, changes light intensity (which we can see) into audio sound that we can hear.

The Sonalight

The Sonalight is an audio oscillator whose frequency of oscillation is dependent upon the intensity of light that impinges on a photocell. Light intensity levels of the Sonalight can be as great as that received on earth from the sun or as weak as light as that from distant stars on a moonless night. This ratio of light intensities is about 100 million and the Sonalight will operate over this range of intensities. Bright sunlight or a nearby 100-watt bulb will produce a frequency of about 6000 to 8000 Hz, while dim starlight or a dark closet will produce about 1 pulse per second. Because of the variation in tolerance of resistors and capacitors the actual upper and lower limits are approximate.

Circuit Analysis

In Fig. 6-1 we see a general circuit for an oscillator which makes use of the unique features of the 555 timer IC chip. This circuit will operate at up to several hundred kilohertz. However, we will use it in the audio-frequency range from a nominal frequency from 1 Hz to about 8 kHz. With the two resistors and single capacitor in the circuit the frequency of oscillation is determined by the equation:

$$f_o = \frac{1.44}{(R1 + 2R2)C1}$$

where,

f_o is the frequency of oscillation in hertz,
R1 and R2 are the values of the resistors in ohms,
C1 is the value of capacitor C1 in farads.

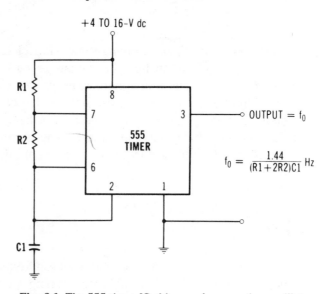

Fig. 6-1. The 555 timer IC chip as a free-running oscillator.

The output is taken from pin 3 of the 555 IC chip and has sufficient power to drive a relay, speaker, or LED. The IC chip operates from a power supply of 4 to 18 volts dc.

In Fig. 6-2 we see the general circuit modified very slightly to use a photocell in place of R2. The photocell resistance will change depending on the intensity of the light that strikes its surface. The photocell (R2) has a resistance of about 10 megohms in the dark and its resistance drops very rapidly under bright light to about 100 ohms. When substituted in the formula for frequency, we find that the frequency will vary from about 1 pulse per second (no light) to a nominal 6000 to 8000 Hz (bright light). Resistor R3 (10 kilohms) has been added to the circuit to hold the upper frequency range to the normal, useful audio-frequency range, a nominal 8 kHz. If R3 were not in the circuit, the upper frequency range in bright sunlight would go up past 18 kHz and could not be heard.

A speaker provides room-level audio output, while an LED is used as a pilot light. The LED will appear to be on constantly once the frequency gets above about 10 to 12 pps. Below that frequency the LED will flash on and off according to the duty cycle. The on-to-off duty cycle is about 60 percent. The current drawn from a 9-volt transistor battery is about 4 mA at 6500 Hz and about 7 mA below 1000 Hz. Therefore, with normal use of a half hour or so per day, the 9-volt battery will last several months.

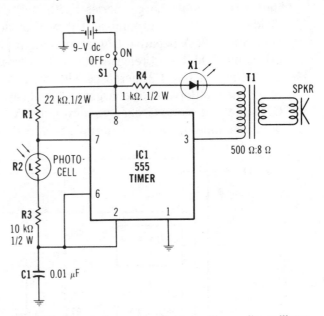

Fig. 6-2. Circuit diagram for light-sensitive audio oscillator.

Component Description

Let us now briefly examine the 555 IC timer and the photoresistor.

555 IC Timer

The 555 timer IC may well be the first IC chip you will work with and it will help to launch you into the amazingly small world of the integrated circuit.

The 555 timer has dozens of uses, the most popular being as an oscillator or timer. Because so much is accomplished inside the chip very few external components are necessary in order for the chip to function in a variety of applications. How you hook it up makes all the difference.

Let's take a look inside the 555 chip, which appears in block-diagram form in Fig. 6-3. The IC chip is one-half the usual 16-pin DIP (dual in-line plastic) package when looking at it from the top. Pin 1 is to the left of a small circle on the top of the package. The internal structure of the 555 chip includes a flip-flop stage, two voltage comparators, a discharge transistor, and a reset transistor. For the sake of discussion we will look at the internal portion of the IC chip with C2 added from pin 7 to ground. The circuit operates in the following manner. The flip-flop normally biases transistor Q1 on so that it conducts and shorts out capacitor C2 to ground. C2, therefore, cannot charge. Series resistors R5, R6, and R7 form a voltage divider from V_{cc} to ground and they provide bias to comparator 1 and comparator 2. A negative trigger applied in pin 2 sets the flip-flop, which drives pin 7 high and releases the short across the external capacitor C2. The voltage across C2 increases with time to $2/3\ V_{cc}$, where comparator 1 resets the flip-flop and discharges external capacitor C2 to $1/3\ V_{cc}$. Capacitor C2 thus charges and discharges between $1/3\ V_{cc}$ and $2/3\ V_{cc}$. The charge and discharge times are set by external resistors R1 and R2 in Fig. 6-1 so that the

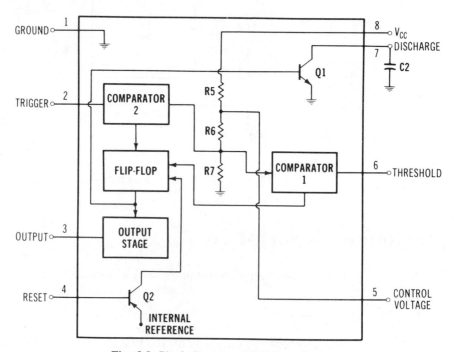

Fig. 6-3. Block diagram of 555 timer IC chip.

frequency of oscillation is determined solely by external components R1, R2, and C2. The frequency is independent of the supply voltage from about 4 to 18 volts dc.

The Photoresistor

In Chapter 4 we discussed in detail the operation of the photoresistor. This device is also known as a light-sensitive resistor, cadmium sulfide (CdS) photocell, and photoconductive device. (See Fig. 6-4.) To give you an idea of the great range of light intensities the photocell can handle, Table 6-1 shows the approximate frequencies produced by different light intensities. These are nominal frequency values produced by average-tolerance resistors and photocells.

Table 6-1. Sonalight Frequency vs. Light Intensity

Scene	Light Intensity (candela/m²)	R2 Value	Frequency (Hz)
Bright Sun	10^4	$100\ \Omega$	7000
Dusk	10^0	$10\ k\Omega$	3000
Starlight	10^{-4}	$10\ M\Omega$	1–2 pps

Fig. 6-4. Cadmium sulfide (CdS) photocell. *Courtesy Radio Shack, Div. of Tandy Corp.*

Construction

As with the other projects the Sonalight can be mounted in any type plastic box or container. Fig. 6-5 shows the Sonalight configured in a plastic case. The power on/off switch is a slide switch and is mounted on the right side of the cabinet so that it can be easily turned on and off with your right thumb. The unit is aimed much like a flashlight, so the photocell is mounted on the end of the plastic box. A hole is drilled to match the diameter of the photocell, and a small plastic-pen section is used as a light funnel. Use a glue or cement to hold the photocell and the plastic tube in place. IC chip printed-circuit card holders are available at most radio supply stores. Table 6-2 is a parts list for the Sonalight.

Table 6-2 Parts List for Sonalight

Item	Description
V1	Transistor Radio Battery, 9-V
C1	Capacitor, 0.01-μF, 15-V dc
IC1	555 Timer IC Chip
X1	Red LED
R1	Resistor, 22-kΩ, $\frac{1}{2}$-W
R2	Photocell, Cadmium Sulfide (CdS)
R3	Resistor, 10-kΩ, $\frac{1}{2}$-W
R4	Resistor, 1-kΩ, $\frac{1}{2}$-W
T1	Transformer, 500 Ω : 8 Ω
SPKR	Speaker, 8-Ω Miniature
S1	Switch, On/Off SPST
Misc.	Plastic Utility Case, 9-V Battery Connector Clip, Hookup Wire, Solder, 16-Pin IC PC Card, etc.

The 16-pin DIP printed-circuit card can be cut in half with a small hacksaw and one-half of it (four pins on each side) used as the printed-circuit card into which the 555 timer is inserted. Solder the leads to the IC chip and the connection wire leads to the copper foil on the printed-circuit card. Use just enough solder to make a good connection. Be sure the solder does not bridge the copper foil. Use epoxy or cement to hold the printed-circuit card in the cabinet and the speaker to the top of the case. Drill a few holes in the case to let the sound out from the speaker.

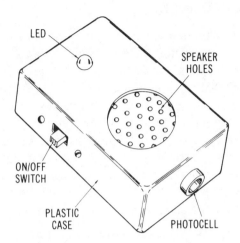

Fig. 6-5. Sonalight mounted in a plastic utility case.

Testing for Operation

Check the wiring carefully before inserting a 9-volt battery in the unit to reassure yourself that you have completed all circuit wiring correctly. Insert a battery in the clip, put the battery in the case, and hold it firmly in place by using styrofoam stuffing cut from egg cartons or the like. Turn the unit on and you should immediately hear a tone from the speaker, and the LED should be on, showing that the audio-frequency current is passing through the LED and the speaker. As you aim the Sonalight at a bright light the tone should increase greatly (a nominal 6 to 10 kHz) and then rapidly decrease when you shut the light off. If it is totally dark in the room, the oscillations should decrease in a few seconds to just one "plop, plop, plop" every second or so. This tells you that the unit is working properly and you are ready for some interesting experiments.

Uses of the Sonalight

The Sonalight is an interesting device; you will find yourself taking it with you wherever you go, to see what experiments you will be able to conduct. Detailed below are a number of experiments or observations that you may perform. Additional refinement of a particular technique might turn an experiment into a worthwhile operational device. Let's take a look-hear!

Intrusion Detection

Intrusion detectors work on basically two different principles—the presence of a new signal or the absence of an old signal. This is true for light beam interruption, ultrasonic (40-kHz) interruption or new presence, microwave (5000-MHz) beam interruption or new presence, magnetic field disturbance, sound vibration due to an intruder walking, electrostatic field disturbance (body capacitance), infrared beam interruption, or infrared body-heat detection. With the Sonalight we will use the quick decrease of light to signal a disturbance or the increase in light to note a change. Set up and aim the Sonalight in such a direction that it receives a certain amount of light from an open window, a distant tennis court that is lighted at night, the hall lights in a shopping mall, or the like.

When a shadow is cast across the Sonalight photocell by a person or object, the frequency will decrease—the amount of decrease depending on the darkness of the shadow. If you are in earshot of the tone from the Sonalight, you will hear a rapid decrease in frequency and then a rapid increase as the shadow passes the photocell. You can also point the photocell at a large bright object, such as a white refrigerator or wall, and have the Sonalight change frequency when someone walks between the white reflecting object and the photocell. This means of alerting you that someone has entered an area is particularly important or useful to a blind person who might be working at a counter or bench, such as a newsstand, as he or she will be able to make use of the light variation rather than by sound alone.

Observe Shadows or Lights

The person with normal sight seldom realizes the marvels of light and shadows and the kinds of music they play. With the Sonalight, you will be able to listen to this music, especially when you ride in a car down a tree-lined highway in the country or ride a bicycle down a tree-shaded lane in a park. Set the Sonalight on the automobile dash pointed in the general direction of the sun or bright area of the sky. The Sonalight will sound at its upper frequency range because of the bright light. When you drive under a group of trees the bright flashes of light will be broken up by bundles of darkness. This will cause the tone to change very rapidly from maximum to near minimum, producing a musical sound during transitions. You will also be able to do this on certain cloudy days when the sun breaks through the clouds, producing a slowly changing series of tones. In England the BBC radio announcers describe this type cloud-cover weather as "Occasional bursts of sunshine!"

At night when you are driving down the street, set the Sonalight on the dash and note the rapid change in tone as you drive under streetlights. When stopping at traffic stoplights you will be able to observe tone changes as a stoplight goes from green to yellow to red and then suddenly to green.

Observe Distant Lightning

The cadmium sulfide (CdS) cell has response above and below that of the human eye. Accordingly it can "see" light that the eye cannot see (though the definition of light covers only electromagnetic energy that the eye can see). This is most evident when it is pointing at distant lightning, where the Sonalight will produce an increase in frequency but you do not notice any flash. The photocell will also catch flashes backscattered from clouds which may be too short in duration for the eye to perceive. At night when you are camping out the Sonalight can be left pointed in the direction from which storms usually arise to signal their arrival when they are 10 to 20 miles away by responding to the distant flashes against the dark night sky.

Find Archery Target in Dark

You can run some simple experiments with a good flashlight, the Sonalight, and a good dark outdoors, around the house or in the park. Aim the flashlight and Sonalight in the same direction and you will be able to pick out the center of an archery target from about 20 feet away by moving the light radar around in an up/down and left/right direction. When the signal frequency jumps up rapidly, you'll know you have scanned across the target. With a little practice you will soon know when you are aiming the "radar" at the target by detecting backscattered light. By the same token, you can place a low-power lamp or flashlight on the target and aim at it with the Sonalight. You'll know the Sonalight is on the target when the frequency is the highest. This is shown in Fig. 6-6.

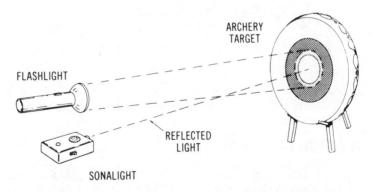

Fig. 6-6. Aiming at archery bull's-eye with flashlight and Sonalight.

Pointing accuracy and sensitivity can be increased considerably by aligning a small, low-power telescope to look into the photocell. You can also use one eyepiece of a pair of binoculars to do the same thing, while you aim through the other eyepiece to let you know you are on target.

TV Receiver

A TV receiver picture has light features which change considerably with time such as light intensity, color, background, action, and technical characteristics. When you listen to an operating TV set, you will hear a buzz of about 30 pulses per second, which is the picture frame rate. In addition, you will hear a complicated frequency change as the video content changes with the picture scene. Lettering will cause the greatest frequency changes as letters have the greatest contrast changes. But remember, you are listening to light intensity changes and not color (frequency) changes when you use the Sonalight. You'll find the TV set an interesting item to listen to.

Awaken at Dawn

Getting up at the crack of dawn is easier said than done. If you need something to waken you when it becomes light, you can use the Sonalight for that. Aim the photocell in the direction the sun will rise and turn the Sonalight on. In total darkness it will pulse about once each second or so but will sound off when the sun comes up or an approaching automobile illuminates your camping area.

Sense Liquid-Level Surface Movement

In certain chemistry experiments and manufacturing processes it is important to know that a liquid surface is still moving or is in a state of agitation. By shining a light at the surface of the liquid which is moving, the reflected light changes greatly in intensity. We will look at this reflected light with the Sonalight and find that it changes frequency rather randomly, just as the surface level does. This action is shown in Fig. 6-7.

Of course, you can use the same general technique to monitor motor rotation, fan movement, gear rotation, or any process that involves movement that must not cease, be it chemical, mechanical, or electromechanical.

Audible Horizontal-Level Device

The bubble level was invented in AD 1625 and it has not been improved much since that time. While not all-electronic in nature the improvement obtainable by using the Sonalight is an interesting innovation since it does permit an accurate sense of horizontal level by using a changing tone to signal a horizontal plane. We will shine a light on the bubble from one side of a level plane and look and listen on the opposite side. The frequency will increase to a maximum when the bubble is centered and decrease when the bubble is either side of the horizontal center or level. Using this audio-frequency level one man can check the level at the center of a beam while he is at one end raising or lowering the beam. He listens for a maximum frequency to indicate a level condition rather than looking at a bubble as has been done for over 300 years.

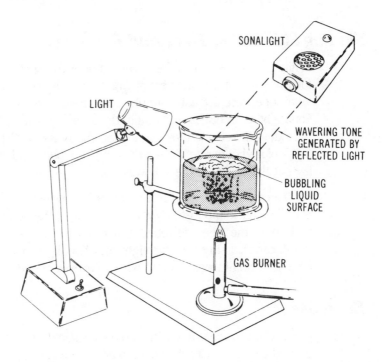

Fig. 6-7. Liquid surface movement detected by light and Sonalight.

Listening to Your Pulse

It is fairly easy to count your heartbeats by feeling lightly on the inside of your wrist for your pulse. More recently, electronic means of taking one's pulse have been available, but they are devices which ordinarily cost $100 to $300. We can use the Sonalight and a light source to sense the beating of the heart.

There are two areas which are usually used to sense the pulsing of the blood—the ear lobe and the fingertip. When the heart beats it sends spurts of blood through the arteries. At the ear lobe and the fingertip, the blood arrives in spurts and the increased volume of the blood can be used to count the pulse.

By shining a light through the ear lobe or the fingertip of the forefinger, and placing the photocell on the opposite side of the ear or finger, you will hear an increase in frequency each time the heart pulses and then a lowering of frequency during heart rest. The forefinger is the easiest finger to use as you place it over the photocell and then hold the Sonalight near a high-intensity desk lamp. You will then hear the sound waver in frequency. Count the number of beats or wavers you hear in 60 seconds and this is the heart pulse rate. A normal heart is about 72 beats per minute when at rest. You might have to work with this technique a bit until you get it to work properly. But it will work, so don't give up.

Finding Sidewalks, Trees and Buildings

While the simple tests to be covered under this section tend to have application for the blind and work best at night when it's dark, they do help to illustrate what one can do with simple equipment. Again, using the flashlight and Sonalight as an optical radar, aim the units at a sidewalk at night. You'll see that the sidewalk reflects more light than the grass or dirt off to either side of the light-colored concrete. Sweep the Sonalight back and forth and you'll be able to lay out the direction the sidewalk goes.

Trees can be located from 5 to 10 feet away by scanning the flashlight and Sonalight back and forth as you walk at night. Even though the bark of a tree is fairly dark the tree will reflect much more light than the dark background in the distance. Using the flashlight and Sonalight, you can easily walk around large and small trees at night with your eyes closed.

Briefcase

If you want to have fun with someone, place the Sonalight inside a briefcase, desk drawer, or closet. Then, when someone opens up the briefcase the Sonalight will suddenly sound out as the frequency increases with the increased light, and the person will jump at least 3 feet!

Police Bubble Machine

The police cars and various other emergency vehicles nowadays are equipped with red, white, or blue rotating and flashing bright lights. At night when they are on they produce a weird optical scene to the eye. When you "listen" to their strange rotating lights, they produce a most unusual pulsating sound that is difficult to describe. By the way, when you listen to them, be sure the police car is passing you, and not stopping you! Conduct your Sonalight experiments in the most inexpensive manner possible.

7 Polarity-Sensing Continuity Tester Using LEDs

This compact solid-state, polarity-sensing, continuity tester is a useful electronic test tool at any workbench. With it you can easily check for shorts, opens, and continuity; you can also sense polarity and condition of diodes, transistors, LEDs, most capacitors, inductors, and any resistive circuit component with a resistance value up to about 30 kilohms.

Operation

The circuit uses two LEDs that operate off alternating voltage and draw current through a common resistor which acts as the current limiter for the LEDs. In a novel circuit arrangement, current is first passed in one direction through the device under test and then in the other direction. If the device being tested is purely resistive, such as a carbon resistor, current will flow through the test resistor in both directions, and both LEDs will light with the same intensity. Each LED in turn will draw current through the series dropping resistor so that the voltage drop across the reverse-polarity, parallel-connected LEDs never exceeds the nominal operating voltage of 1.6 to 2.0 volts dc. The forward conduction of one LED protects the other LED from reverse voltage conduction during alternate cycles of the ac voltage. This occurs 60 times a second, and to the eye the LEDs appear to remain on continuously. The test device is very simple but provides a wealth of information about the condition of almost any electronic component. The continuity tester is shown in Fig. 7-1 mounted in a plastic case.

Fig. 7-1. Continuity tester uses direction of current for a test condition.

Circuit Analysis

The circuit for the continuity tester is shown in Fig. 7-2. It operates off 120-V ac house current and uses a transformer to provide a nominal 12.6-V ac, 60 Hz, which also provides isolation from the ac supply line. The series current-limiting resistor, R1, limits the amount of current that can flow through the LEDs when they are conducting. Most LEDs have a nominal current rating of 20 mA, and this value of current is used to determine what the value of R should be to limit the voltage drop across the LEDs to a nominal 1.6 volts for red LEDs or 2.0 volts for green LEDs. The peak voltage across the secondary winding is equal to 12.6-V ac × 1.414, or 17.8-V ac. Since the LED is to have a 1.6-volt drop across it, the resistor must drop the remainder, or 17.8 − 1.6 = 16.2 volts. We then compute the value for R1 necessary to drop 16.2 volts at 20 mA. We have, according to Ohm's law:

$$R1 = \frac{E}{I}$$

where,
    R1 is the value of resistor R1 in ohms,
    E is the voltage to be dropped across resistor R1,
    I is the current through resistor R1 in amperes.

Accordingly we have

$$R1 = \frac{E}{I}$$

$$= \frac{16.2}{20 \times 10^{-3}}$$

$$= 810 \text{ ohms}$$

is the resistance. Since 1 kΩ is the nearest standard value, we have used this value in Fig. 7-2.

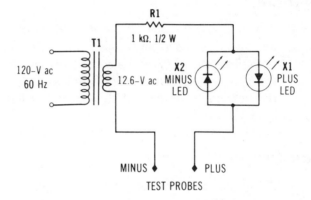

Fig. 7-2. Continuity tester uses a minimum of parts.

The two LEDs can be red or green but they should be the same color and chosen so that they have about the same brilliance. This is so that a comparison of brightness can be made when checking diodes and the like. An LED is now packaged which has a red and green LED in the same epoxy envelope. The LEDs are connected in reverse-polarity parallel and are well suited for this project except for the possible difference in brightness of the red and green colors to the eye. Green is at the peak of the response of the human eye (550 nm) while the red LED shines at a frequency (650 nm) where the response is only 5 to 10 percent of its maximum (green). Thus, for the same amount of energy the green LED will appear to be many times brighter than the red LED. An LED should never be operated from an ac voltage source without having a second LED or diode in reverse-polarity parallel with it, as it only has a reverse voltage rating of a nominal 3 volts. Breakdown of the LED can occur if this voltage is exceeded. An ordinary signal diode can take a nominal 50 to 100 volts before damage occurs.

A pair of test leads 2 to 3 feet long with insulated alligator clips are used for making connection to items to be tested. When the clip leads are shorted to each other, the LEDs will glow at their maximum brightness. This is the condition for checking for a short circuit—both LEDs lit at maximum brightness. Since both LEDs are lit, current is flowing through the short in each half cycle of the ac sine wave.

Testing Bidirectional Devices

The continuity tester is very useful for testing many bidirectional devices.

Light Bulbs

A cold light bulb will have 5 to 10 ohms of resistance, so when the probes are placed across its terminals, both LEDs should light as current can flow in both directions. This will also be true for pilot lights, flashlight bulbs, high-intensity lamps, automobile headlights, and the like. If there is an open in the light bulb, neither of the LEDs will light as there is not a complete circuit.

Transformer

A transformer, inductor, solenoid, or relay winding consists of a number of turns of wire that will ordinarily have a dc resistance of a few tenths to a few hundredths of ohms. When one of these is connected across the test probes, current will flow through the LEDs in both directions and both LEDs will light. You will then know that the winding is good. In the same manner you can test speakers, earphones, and piezosounders. At this time, however, you will also hear a 60-Hz hum and both LEDs will light, showing that the device being tested is good.

Photoresistors

As discussed earlier in other chapters the resistance of a photocell, or photo-resistor, varies greatly, from a few hundred ohms to 1 to 10 megohms, depending on the amount of light striking its surface. When we connect the test probes to the photoresistor, both LEDs will light as the photoresistor is a bidirectional device. As you cover the photocell with your hand, however, the LEDs will dim greatly as the resistance increases up to several hundred thousands of ohms. This will show you that the photocell is working properly.

The Resistor

A resistor will pass current through it in both directions, so when one is connected at the test probes both LEDs should light with the same intensity. Using different values of resistors place each one across the probes and note the resistance from the color code when the LEDs are very dim. This will be on the order of 20 to 30 kilohms, depending on the type LEDs you have used. If you remember the brilliance of the LEDs at this high resistance it will help you to judge the front-to-back ratios of diodes.

Testing Unidirectional Devices

In this section we will discuss the testing of unidirectional devices and learn that the polarity tester is very useful in this mode.

The Diode

The simple diode is a unidirectional, solid-state device which we covered at length in Chapter 1. We learned that electrons will travel through a wire, or solid-state material, from a negative terminal, such as a cathode, to a positive terminal, such as an anode, of a diode. Let's connect a diode to the continuity/polarity tester as we see in Fig. 7-3. When point A is positive, electrons will be able to travel from the negative terminal of the 12.6-volt winding through the negative terminal of the tester to the negative terminal (cathode) of the diode, through the diode to the anode of the diode, to the positive-marked terminal of the tester, to the cathode of the LED X1, to the anode of X1, through the 810-ohm resistor to positive supply voltage terminal A of the 12.6-volt winding.

When current passes through X1 in the direction indicated, X1 lights, but no current can flow through the diode in a reverse direction, so X2 remains off. If we were to reverse the position of the test diode, current would flow through the diode and X2 would light and X1 would be out. Note, however, that we did not have to reverse the position of the diode, or the probes, but the ac current did it for us. Accordingly, we wire up the positive (red, or plus) probe so that X1 lights when the probe is connected to a positive terminal or anode of a test diode, LED, or transistor. The negative terminal of the tester, of course, must be connected to the cathode, or negative, terminal of the same device. However, when the positive terminal lead is connected to a cathode (negative) terminal of a diode, the minus LED will light, which tells you that the positive probe is connected to a negative (or cathode) terminal. You don't have to reverse the test leads to make the determination of the polarity of the device under test. If

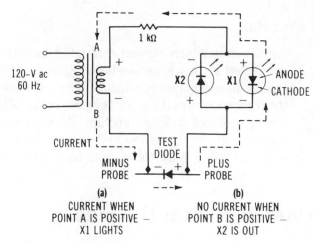

Fig. 7-3. Continuity/polarity tester connected to diode with forward conduction of current (a) and no current with reverse conduction (b).

you were to use an ohmmeter to make such a test you could make an incorrect guess 50 percent of the time, and have to reverse the ohmmeter probes.

One other observation is to be made from the brilliance of the LEDs. Their brilliance is an indication of the front-to-back ratio of the diode. A diode with a good front-to-back ratio would have one of the LEDs really bright while the other LED might be out completely or just slightly lit. If both LEDs are lit, the diode is obviously shorted as current flows in both directions in the diode and through each LED. If neither LED is lit, obviously there is no current through the diode. So with one connection of a pair of test probes to the test device, we can examine a diode in good depth. This is depicted in Fig. 7-4. A simple test device such as this is very worthwhile on a test bench or a production line as a jig can be used for inserting the device to be tested and many of them can be tested per unit time.

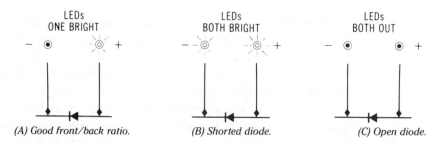

Fig. 7-4. Three indications given in testing a diode for continuity.

The Solar Cell

A selenium solar cell has about 1000 to 1500 ohms of resistance, with a poorly defined front-to-back ratio. That is, it has about the same resistance in either direction, and this is affected by the amount of light striking it during the resistance measurement. Both LEDs will light when the test probes are placed across the solar cell leads (showing there is little front-to-back ratio) but the LED brightness should be somewhat less than when the test leads are shorted. So always short the test leads together to get an idea of the relative brightness of the LEDs and then place the leads across the device to be tested. If the device is installed in a circuit, however, be sure that there is no voltage applied to the circuit to be tested, and also that you are testing just the device and not some other part or portion of the circuit which is connected across the device.

Construction

The parts necessary to build the continuity and polarity tester will easily fit in a small black plastic case. Drill two small holes in the upper portion of the case top for press insertion of the LEDs as well as two holes near the bottom for the two test leads. Refer to Fig. 7-1 for an idea of the relative placement of the holes. Drill one hole in the side of the plastic case near the bottom for the ac line cord to enter. Tie a knot in the case end of the ac cord so that the cord will not pull out and exert pressure on the wired circuit. Test leads with insulated alligator clips can be found at the local electronics stores. Use a red, green, or yellow colored lead for the positive (plus) connection and a black lead for the negative (minus) lead. Also, label the LEDs with a labeling machine with a *P* and *M* to remind you that when the *P* LED is lit, the positive probe or lead is connected to the anode, or positive terminal, of the device under test. If the *M* or minus LED is lit, the positive probe is connected to the negative, or cathode, of the device under test. The parts list for the continuity tester is shown in Table 7-1.

Table 7-1. Parts List for Continuity/Polarity Tester

Item	Description
X1, X2	Red or Green LED, 20-mA Current
R1	Resistor, 1-kΩ, $1/2$-W
T1	Transformer, Primary, 120-V ac; Secondary, 12.6-V ac, Radio Shack 273-1385 or equivalent
Case	Plastic Case, Radio Shack 270-231 or Equivalent
Test Leads	Test Leads 24 to 36 inches (60 to 90 cm) with Alligator Clips, Radio Shack 278-001 or equivalent
AC Cord	AC Line Cord, Radio Shack 278-1255 or Equivalent
Misc.	Hookup Wire, Solder, Screws, etc.

Testing for Proper Operation

After all components have been mounted, wire the circuit as shown in the schematic (Fig. 7-2). Mark off on the diagram that portion of the circuit that you have completed. Use two small nuts and bolts to hold the transformer to one side of the plastic case. Solder all connections and tape the ac cord where the leads connect to the transformer. This circuit uses voltages which are dangerous, so exercise caution when the unit is plugged into the ac outlet and the case is open. The 120-V ac will give you quite a zap if you get across the exposed leads, so be careful. Always unplug the unit when you are working on the inside of the case. Remember, 120-V ac can kill, so have great respect for it!

Plug the ac cord into the ac outlet, touch the two test probes together, and both LEDs should light. Now try a diode across the probes and reverse the position of the diode to see that the LEDs light properly as labeled. If both LEDs light in one direction of the diode but not the other, the LEDs are wired in parallel and not reverse-polarity parallel. Make the wiring corrections as necessary.

Additional Uses

Continuity for resistance values up to about 20 kilohms can be observed with the glow of the LEDs. Capacitors with values larger than about 0.05 μF can be tested with a faint glow of both LEDs, showing the capacitor is good at 60 Hz. To calculate the value of impedance, use the expression:

$$X_c = \frac{1}{2\pi fC}$$

where,
    X_c is the capacitive reactance in ohms,
    f is the frequency,
    C is the capacitance value in farads,
    π is the constant 3.1416.

Thus for a 0.05-μF capacitor at 60 Hz,

$$X_c = \frac{1}{2 \times 3.1416 \times 60 \times .05 \times 10^{-6}}$$

$$= 5305 \text{ ohms}$$

The brightness of the LEDs produced by the 0.05-μF capacitor can be compared to the brightness for a 5305-ohm resistor.

Speakers and headphones can be tested for continuity; they will also produce a soft 60-Hz hum, the frequency of the applied ac.

In normal operation the continuity/polarity tester draws approximately 0.03 watt. Thus, it can be left plugged in indefinitely so that it is always ready to test those many items for which a continuity test is just what the technician ordered.

A Test Set and Power Supply for LEDs

The continuity/polarity tester you have built is a very powerful and unique device for testing other LEDs. It not only tests them but it serves as a shortproof power supply for LEDs.

Visible LEDs

You can take an unknown type LED with unknown color and unknown polarity and connect the probes of the tester across its leads. If the LED is good, it, as well as one of the tester LEDs, will light. It makes no difference how you have connected the LED to the test probes because you can identify the LED anode and cathode without having to look for a longer or shorter lead or a flat edge on the epoxy base of the LED. You can tell something about its brightness by how bright it is compared to other LEDs or the brightness of the tester LEDs. If you short out the test LED, the LED will go out and both tester LEDs will now be lit. So you have no need of fearing anything if you short out the tester leads.

IR LEDs

A property of an infrared (IR) LED is that you cannot see its emission. Its emission wavelength is below the response of the human eye, near 950 nm, while the response of the eye is from a nominal 450 nm (blue) to 750 nm (red). So how do you know you have a good IR LED without hooking it up to some elaborate system and making the system work in order to test the IR LED? Well, you and the author know we will test the IR LED. We'll put it across the polarity tester and look for the common symptoms of a good LED or diode. If one tester LED lights, the IR LED is good (and we have identified the anode and cathode leads, which is especially useful if you don't have the technical data!). If both tester LEDs light, the IR LED is shorted, and if neither tester LED lights, the IR LED is open. So we have not seen any light from the IR LED but we know it is good!

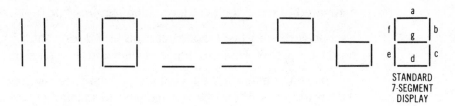

Fig. 7-5. Unusual designs you can make with a seven-segment display.

Digital Displays

A seven-segment LED digital display can easily be checked for proper operation and to determine if it is common anode or common cathode by using the polarity tester. Using what you have learned so far you can rapidly identify the common terminals and then check each of the seven segments and the decimal point to see that they are all in order. You can also easily see which leads you need to connect to come up with unusual designs such as is shown in Fig. 7-5. You can parallel as many of the segments as you want and light them from the two leads from the polarity tester. You will find your test instrument one of the most useful and frequently employed test devices that you have on your bench!

8 A Light-Intensity-Level Wheatstone Bridge

A sensitive photocell light-level sensor can be built using LEDs as indicators to sense when the light level increases or decreases from a set value. Its circuit operation is simple and we will build a unit into a handheld package that is battery operated.

Operation

Fig. 8-1 shows the unit packaged in a small plastic case. The design of the circuit is such that when we point the photocell unit at a light source of a certain level (such as at an open window or near a lamp) one of the LEDs will light. By zeroing the bridge used in the circuit, both of the LEDs will go out, showing that the bridge is balanced. This means there is no current though either LED in either direction. When we unbalance the bridge by bringing it near a dimmer or brighter light source one of the LEDs will light and one will remain out. When we revert to our original light intensity, however, we again have a balanced light condition and both LEDs will be out.

Circuit Analysis

The light-level sensor bridge uses the same circuit as developed by Sir Charles Wheatstone (1802–1875) of England over 100 years ago. This circuit has remained virtually unchanged over the past century and has been used

Fig. 8-1. Light-intensity bridge in a handheld package.

by numerous electrical and electronics engineers, technicians, teachers, and students over the years. One of its primary uses is to measure the value of unknown resistors accurately. It can also be used on alternating current and will operate at radio frequencies to determine unknown values of inductance and capacitance. In this project we will use it at light frequencies by means of a photocell.

Fig. 8-2 shows the schematic; we see that it is a very basic and simple circuit. We have replaced several components, namely the ammeter, or current meter, with two reverse-polarity parallel-connected LEDs, and one of the resistors with a photoresistor, which we discussed earlier. Fig. 8-3A shows the usual schematic for the Wheatstone bridge, which is not simple to analyze as it appears in an awkward component arrangement. When we arrange it in a more easily understood circuit we see that we can quickly analyze and understand its operation. Looking at Fig. 8-3B we see that the resistance values have been chosen so that the voltage drops from points *BD* and *AD* are equal and that the voltage drops from points *BC* and *AC* are equal. Therefore, points *A* and *B* relative to *D* have the same potential and there is no voltage difference between points *A* and *B*. If there is no voltage difference there can be no current through the meter, and we can say that the current through the meter is zero, or $I_{\text{METER}} = 0$. The meter will, therefore, indicate in the middle of the center-zero scale, showing that there is no current through it. Summarizing what we have covered, if we apply 10 volts across point *C* to point *D* we can see that because the voltage divides evenly according to the amount of resistance in the circuit we will have 5

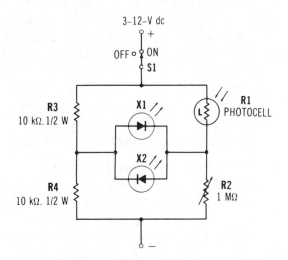

Fig. 8-2. The light-sensitive bridge is simple and sensitive.

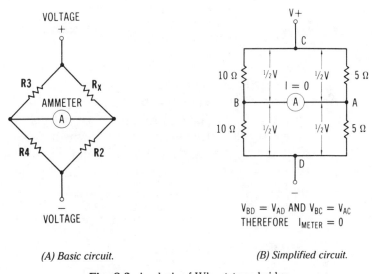

(A) Basic circuit.

(B) Simplified circuit.

Fig. 8-3. Analysis of Wheatstone bridge.

volts from point A to D and 5 volts from point A to C. Likewise, we will have 5 volts from point B to D, and 5 volts from point C to B. Therefore, there is no voltage difference from point A to B because both points are at plus 5 volts with respect to point D. Thus, there will be no current from point A to B or from point B to A.

Let us now unbalance the bridge, that is, vary the resistance R_x in Fig. 8-3A or change the 5-ohm resistor (from point A to C in Fig. 8-3B) to a value of 3 ohms. When we look at the voltages across the 10-ohm resistors they remain the same (5 volts), but when we look at the voltage across the 5-ohm resistor from point A to D we find that the voltage drop has increased from 5 volts to 6.25 volts because the total resistance of leg DAC is 8 ohms, not 10 ohms. So the voltage drop across the remaining 5-ohm resistor is $5/8$ of 10 volts, or 6.25 volts. This means that point A to D now has 6.25 volts, while from point B to D we still have 5 volts. Thus point A is 1.25 volts more positive than point B, and current will flow from point B to point A. The meter pointer would thus move to the right.

If we were to reverse the situation and make the R_x value of the 5-ohm resistor 8 ohms instead of 5 ohms, point A would have less than 5 volts and current would flow through the meter in the opposite direction. We then have learned that any time we unbalance the bridge circuit, current will flow. And thus it is for our bridge circuit which uses a photocell and LEDs. We use a change in light intensity to unbalance the bridge and we use the LEDs to show direction of current.

Construction

The light-balance bridge is built into a plastic box with components positioned as illustrated in Fig. 8-1. As was done with the Sonalight, a piece of plastic writing-pen barrel about $3/4$ inch (1.9 cm) is inserted in the end of the case to act as a light concentrator for the photocell. Hold it in place with epoxy or glue. The parts list is shown in Table 8-1. An on/off switch is mounted on the right-hand or left-hand side of the case so it can be easily turned on and off with the thumb and unit pointed at a light source like a flashlight.

Table 8-1. Parts List for Light Intensity Bridge

Item	Description
V1	9-V Transistor Radio Battery
X1, X2	Two Red or Green LEDs
R1	Photocell, Cadmium Sulfide (CdS)
R2	1-MΩ Variable Resistor
R3, R4	10-kΩ, $1/2$-W Resistor
S1	SPST Slide Switch
Case	Plastic Case, Radio Shack 270-230 or Equivalent
Misc.	Hookup Wire, Solder, 9-Volt Battery Connector Clip, etc.

Uses of the Bridge

The light-intensity Wheatstone bridge can be used for a number of things and some of them are detailed in the following section.

Light-Intensity Sensing

The intensity of two light sources can be compared by aiming the photocell at a lamp or light source and nulling the bridge with the adjusting control until both of the LEDs are out or have the same intensity. With a perfect balance, both LEDs should be out as there cannot be any current through either of the LEDs. Now, knowing the null is still set at the same level, we move to another light source and increase the lamp intensity (for example, with a lamp dimmer control) until both LEDs are again out. At that time the light intensities of the two sources will be the same. We can make this measurement even though we move from one location to another (a different room, for example) or do the measurement the next day.

Azimuth Sensing to Light Source

We can use the intensity bridge as a light compass, where the LEDs indicate by intensity changes the azimuth and elevation to a light source. Aim the compass at the light source and null the LEDs until they both are out. When you swing the device to the left or right of the source, one of the other LEDs will light, depending on the direction of swing. You can thus swing the device back and forth in azimuth and up and down in elevation. When the LEDs are kept nulled in both planes the photocell is aimed at the light source. This is shown in Fig. 8-4.

Distance Sensing to Light Source

Let's now use the device as a detector of the distance to a light source. It's not that we measure the distance to a light source but that we can tell when we are at the *same distance* to the source. If you are 20 feet (6.09 meters) from a light source, null the bridge until both LEDs are out when the device is pointed at the source. Now, if you move back to a distance of 25 feet (7.62 m) from the light, one of the LEDs will light, as the bridge is now unbalanced. Now if we quickly walk to a point 15 feet (4.57 m) from the light source, the other LED will be lit because we are now looking at a brighter light than when at 20 feet. However, if we now back up to the point where both LEDs are out it means we will be back at the 20-foot position. This means the bridge is again balanced and we will be back at the 20-foot spot within just a few inches of where we started. Refer to Fig. 8-5 for details.

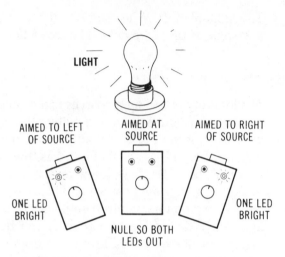

Fig. 8-4. Azimuth sensing on a light source.

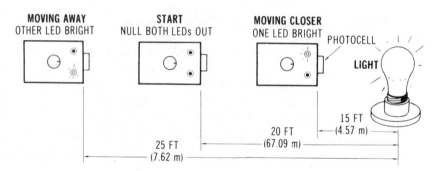

Fig. 8-5. Distance sensing to a light source.

Other Uses

As you gain experience with the light balance bridge you will be thinking of additional uses for the device. These include functions such as sorting, counting, intrusion alarm, position sensor, and edge monitoring device. When optical couplers replace the LEDs, the circuit will provide tracking information for a light tracker to seek out and track a moving light source. In a simple manner you can track the sun, moon, or other bright object, such as an airplane or automobile.

9 An Audio Continuity and Voltage Tester: The Testone

In this chapter we will discuss a continuity and voltage tester which makes use of a sense other than the eye—the ear. In the audio continuity and voltage tester which we will call *Testone* we will use a constant audio tone to tell us of the condition of a circuit under test.

Background

An early test instrument used in the telephone industry for checking continuity between wires was a buzzer and a battery. This test device has not changed much down through the years, so that even today in multimillion dollar industries you may see the buzzer and battery in use, held together with electrical tape. The buzzers were crude but provided the information needed: Was there continuity in the circuit? If there was a buzz it was the correct pair of wires; if not, go on to another pair of wires until a buzz is heard.

Operation

The Testone is a very handy device which uses a simple circuit to serve several purposes. In order to observe continuity using an audio tone we employ a battery, a tone generator (piezosounder), and some test leads.

Fig. 9-1 shows the Testone packaged in a plastic case, with test leads and alligator clips so that one can clip onto the device to be tested.

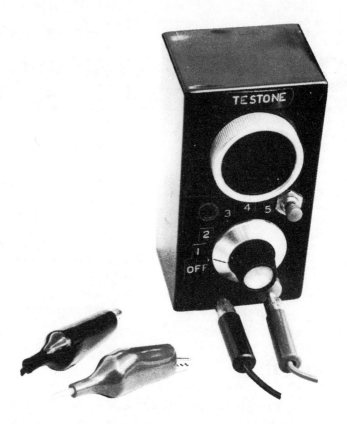

Fig. 9-1. The Testone mounted in a handheld plastic case.

Continuity Testing

In order to test continuity in a circuit we must drive current through the circuit and into a meter or other indicating device. In the Testone we make use of a tone that is produced by a piezosounder. So we listen to the tone with our ears rather than look at a meter with our eyes.

Voltage Testing

In the voltage mode of the Testone we make use of the basic tone-producing device—the piezosounder. The only precaution that we must exercise is to operate or test only circuits that do not exceed the piezo-sounder operating voltage.

Circuit Analysis

The circuit for the Testone is shown in Fig. 9-2. A selector switch is used to select the functions to be used, such as Off, Continuity, and Voltage. "Off" is switch position 1 and "Continuity" is provided by switch position 2.

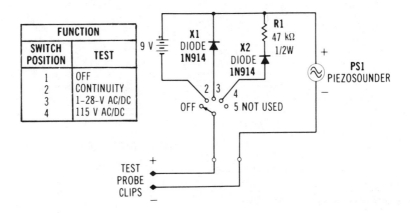

FUNCTION		
SWITCH POSITION	TEST	
1	OFF	
2	CONTINUITY	
3	1-28-V AC/DC	
4	115 V AC/DC	

Fig. 9-2. Circuit diagram for Testone audio continuity and voltage tester.

Continuity

In position 2 the 9-volt transistor battery is placed in the circuit. Note that the positive terminal of the piezosounder is connected to the positive terminal of the battery. The negative terminal of the piezosounder is connected to the negative terminal of the battery through the test probes and the device to be tested. When the probe tips are not in contact, there is no sound because the resistance is infinite. When the probes are touched together there is continuity (no resistance) and the sounder will sound out. These are the two extremes for the resistance or continuity tests of the device. The simplified circuit for continuity tests is shown in Fig. 9-3A.

Voltage Testing

When the Testone switch is turned to position 3 the battery is removed from the circuit, and we switch to the voltage testing mode (Fig. 9-3B). We can measure voltage from 1 to 28 volts ac or dc through use of a 1N914 diode to rectify the ac voltage so that only dc is applied to the piezosounder. When we switch to position 4 we place a 47-kilohm resistor in the circuit in series with the 1N914 diode. This allows us to measure 120-V ac or dc. Notice, in Fig. 9-2 a 5-position switch was used. Position 5 then, essentially, becomes another "Off" position. If you cannot locate the 4-position rotary switch specified in the parts list, substitute a 5- or 6-position switch and leave the remaining positions unconnected.

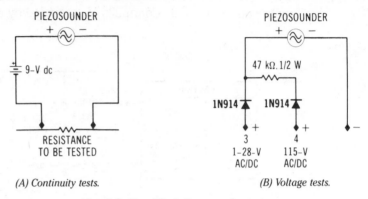

(A) Continuity tests.   *(B) Voltage tests.*

Fig. 9-3. Simplified diagrams for tests.

Construction

The Testone is mounted in a small plastic case with the parts arranged as shown in Fig. 9-1. Cut a hole in the plastic case for the piezosounder by drilling a number of small holes around a circle large enough to mount the sounder. If you use a piezosounder from Radio Shack or other manufacturer, drill a ½ -inch-diameter hole, or several smaller holes, in the center of the case. Mount the sounder to the case with glue, epoxy, or the washer screw supplied. Use point-to-point wiring to connect the various components together. Drill two small holes in the lower front of the case for the two test probe wires. As you wire part of the circuit, mark on the circuit diagram that portion of the circuit that you have completed. Label the switch positions on the front of the case.

The parts list for the Testone is shown in Table 9-1. The Mallory Sonalert (piezosounder) SC628 is available at most industrial and electronics supply stores, while the Radio Shack sounder is readily available at your local store.

Operational Uses

The Testone is very handy for testing a number of devices for continuity and resistance up to about 50 kilohms.

Table 9-1. Parts List for Testone

Item	Description
V1	9-V Transistor Radio Battery
X1	Diode, 1N914 or Equivalent
Case	Plastic Case, Radio Shack 270-231 or Equivalent
PS1	Piezosounder, Mallory SC628 or Radio Shack 273-060 or equivalent
R1	Resistor, 47-kΩ, ½-W
S1	Switch, 4-Position Rotary (see text)
Test Probes	Test leads, 36 inches (90 cm) Long with Alligator Clips
Misc.	Hookup Wire, Solder, etc.

Bench Tests

In testing, a short gives maximum sound, while an open gives no sound. From 20 to 50 kilohms the sound will be very weak and you can determine approximately various resistor values by noting the sound levels produced. Bidirectional devices will produce a tone no matter how they are connected to the test probes, as they have no polarity. Such devices are light bulbs, flashlight bulbs, resistors, transformer windings, solenoids, relay windings and contacts, speakers, earphones, coil windings, and the like.

As well as testing for continuity it is important to test to see whether something is shorted. That is, while you might not be able to measure or test for a high resistance it is important to know that something is not shorted. Such items as insulators, tuning capacitors, open relay contacts, spark plugs and the like should not indicate continuity. Capacitors of 100 μF or larger are interesting to test. When you test them for continuity the battery in the Testone will charge the capacitor, and the tone will be heard as the capacitor is being charged. The tone will last for 2 to 10 seconds if the capacitor is 100 to 1000 μF or so. As the capacitor becomes fully charged the tone will gradually die out. If now you reverse the leads connected to the capacitor and switch the Testone to position 3 (1–28 V dc), the tone will sound again for a number of seconds as the capacitor discharges through the resistance of the Testone. The capacitor now acts as a battery, and when it is fully discharged the tone will stop. From these tests you can tell the condition of the larger-size capacitors. The smaller capacitors, however, will just produce a slight "ping" sound when they are connected in this way to the Testone.

The polarity of the probes must be observed when measuring or testing voltages. The Testone uses a red (green or yellow) probe for the positive terminal and a black lead for the negative terminal. Therefore, when you are charging a capacitor, connect the negative probe of the Testone to the positive terminal of the capacitor and the positive probe of the Testone to the negative probe of the capacitor. Then when you switch to the voltage testing part of the Testone (switch position 3), the charged capacitor acts as a battery and you connect the positive terminal of the Testone to the positive terminal of the capacitor.

Intermittent electrical connections are difficult to detect and locate, but the Testone will aid you in locating them. By listening to a tone (or break in tone) as you wiggle a lead or connection you will quickly know when you have located the cause of the malfunction. If you use an ohmmeter you have to keep glancing at the meter each time you wiggle the connection to see if you have affected the circuit.

Diodes and LEDs

A diode and an LED will produce the same effect when testing them with the Testone. They will conduct in just one direction and not the other. Because we do not have two LEDs to tell us of the circuit condition as we did in Chapter 3 we must reverse connections to the device under test. So always test the diode and LED in both directions. If they are good the Testone will sound in just one direction. A good LED will glow slightly when the Testone sounds out. A correct connection is made when the positive (red) lead of the Testone is connected to the negative (cathode) of the diode or LED. The negative lead of the Testone is thus connected to the anode of the diode or LED.

Bipolar transistors can be tested in this manner, though it does take a bit more of connection juggling than was done for the diode. It is preferable to test the transistor out of the circuit so other components will not influence the results. Since a bipolar transistor can be thought of as two pn junctions we can test them in that manner. Fig. 9-4 shows the connection to be made to a pnp and npn transistor. A properly operating transistor will produce a tone for the four circuit arrangements shown (two pnp, two npn). Since the Testone is a low-current test instrument it is safe to use on low-current solid-state devices.

Automobiles

The Testone is very useful on an automobile, especially in the voltage test connection mode. Use a pin or needle to pierce the insulation when looking for a certain wire on an automobile or truck, trailer, or RV. Connect the Testone positive lead to the pin, and the negative lead of the Testone to negative or ground of the vehicle. Then activate the circuit you are looking for to see if the Testone sounds out as it should for the correct wire. If it doesn't sound out try another wire. In this manner you can look for a circuit

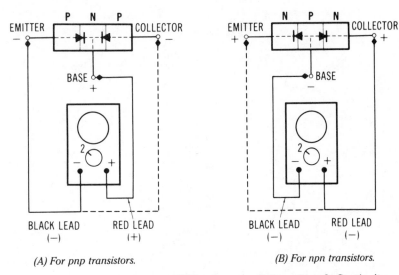

(A) For pnp transistors.   *(B) For npn transistors.*

Fig. 9-4. Testing transistors with Testone on switch position 2, Continuity.

in the trunk of a car while operating the ignition key, gearshift, brakes, horn, windshield wiper, and the like, from the front seat. Since you are listening for a sound rather than trying to read a meter, it is much faster to find a hidden wire while sitting on your head, upside down, by using your ears rather than your eyes. The 12- to 28-volt dc feature of the Testone means you can use it to work on aircraft as well as automobiles. With the diodes provided in the circuit you can also use the Testone on 60- and 400-Hz aircraft alternators. The Testone is, indeed, a useful test device for your bench or toolbox.

10 The LED in Motion

The stop action of the LED display of a pocket calculator was discussed briefly in Chapter 2. Now we will discuss a unique use for the LED, a use that enables us to have a new display device—the moving LED. The moving LED display takes advantage of a unique capability of the LED, i.e., it can be switched on and off very rapidly, in a matter of tens of nanoseconds. This feature, fast switching, when coupled with rapid movement of the LED, opens up a new visual display application area. Applications can be found in logic probes, analog displays for music and voice, time synchronization, frequency measurement and synchronization, spectrum analysis, voice recognition, synchro position resolvers, sonar displays, and multicolor dot-matrix displays. In this chapter we will discuss some of the most easily applied uses of the moving LED.

DC Operation of the LED

We saw in the use of the Probevolt (Chapter 2) and the LED voltage indicator (Chapter 3) that when a dc voltage was applied to the LED it came on and displayed a constant light-intensity output. When we moved the LED back and forth rapidly it produced a bright, steady streak of red or green light. That is, it contained no information in the space, or movement, domain.

AC Voltage Operation

When an ac voltage or pulsating dc voltage is applied to an LED, it will turn on and off, its light intensity varying in an analog manner following the applied voltage. An LED with 60-Hz voltage applied to it will appear to be on continuously, as 60 Hz is above the 10 to 12 "cycles-per-second" flicker rate of the human eye. However, when the LED is moved back and forth rapidly or rotated in a circle as was done in the old *flying spot* scanner of early TV days (before the picture tube), the LED is seen to strobe out a dashed line or circle as shown in Fig. 10-1. A simple circuit for supplying voltage to drive the LEDs is shown in Fig. 10-2. In this circuit the LEDs protect each other from reverse-polarity voltage. While very simple in nature the experiments to be described will be unique and you should find them interesting.

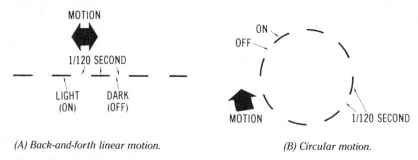

(A) Back-and-forth linear motion.   (B) Circular motion.

Fig. 10-1. When pulsating ac (60 Hz) is applied to a moving LED, broken lines of light are formed.

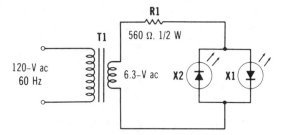

Fig. 10-2. Circuit used to operate LEDs from 60-Hz power.

Bouncing Tennis-Ball Effect

A simple "bouncing tennis ball" experiment can be performed by using an electric saber saw, a lamp dimmer, and a single LED. Mount the LED on the end of the blade of the saw and hold it in place using electrical tape. Wires 1 to 2 feet in length connect the LED to the rest of the circuit shown in Fig. 10-2 (in this case using one LED instead of two) and they provide 60-Hz power to the LEDs. The lamp dimmer is mounted in a small metal or wood box and has an ac outlet as shown in Fig. 10-3. Its ac cord is plugged into a wall outlet. The lamp dimmer can vary the voltage applied to the saber saw from where the blade barely moves back and forth to full power. The dimmer can also be used for other purposes, so it is a valuable device to have around as it can control power to a 600-watt load. Other uses are for a film slide projector so you can dim the lamp to lengthen lamp life, a fish aquarium bubble-machine speed control, Christmas-tree lamp dimmer, and lamp dimmer for a floor lamp for use as a nightlight.

Fig. 10-3. Lamp dimmer installed in metal box with lamp outlet.

The manner in which the LED is mounted on the saber saw blade and circuit arrangement is shown in Fig. 10-4. The saber saw blade will probably have a ½-inch stroke. Plug the LED circuit in and the LED on the saw will light each half-cycle of the 60-Hz power. Start the saber saw and increase its speed. As the speed of the saw is increased you will see that the LED is indeed being turned off and on by the 60-Hz power. Soon the speed of the saw will be a multiple of the 60 Hz powering the LED and it will appear to bounce back and forth like a tennis ball between racket and a wall. Darken the room to help dramatize the effect. You can make the ball stand still in the middle of the stroke, or bounce slowly back and forth between each end, merely by changing the speed of the saber saw. This simple demonstration is to show what you can do with the moving LED. The eye gives the illusion of continuous motion and fixed position of the LED.

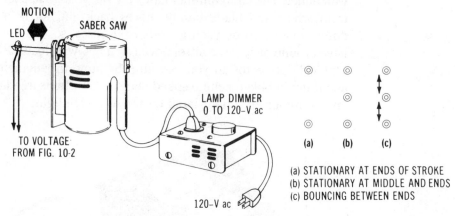

(a) STATIONARY AT ENDS OF STROKE
(b) STATIONARY AT MIDDLE AND ENDS
(c) BOUNCING BETWEEN ENDS

(A) Arrangement of equipment.   *(B) LED as viewed from top as speed varies.*

Fig. 10-4. Circuit for demonstrating moving LED creates unusual visual effects as saw speed is varied with lamp dimmer.

The Rotating LED or SpinLED

A more ambitious project that uses the moving LED involves the use of slip rings to provide power to the LEDs as they are rotating. Fig. 10-5 shows a general layout for a small electric motor, a "color" wheel mounted with LEDs, and slip rings which provide audio power from a radio or stereo source to the LED. Because the two LEDs in each branch are connected in reverse-polarity parallel one will light on the positive swings of the audio voltage and the other LED will light on the negative swings of the audio voltage. The common dropping resistor limits current for both LEDs. A number of LEDs can be placed on the disk so that circles of different diameter will be strobed out as the disk spins.

When music or voices are observed on the spinning LEDs the lower frequencies will strobe out long arcs and the high frequencies will strobe out many short arcs. When there is no speech the display is dark. The display is really dynamic and finds application in home high-fidelity stereo. Through use of appropriate slip ring pairs it is possible to display separate left-right stereo channels. By means of standard audio filters the low, middle, and high frequencies can be separately displayed using red, green, or yellow LEDs.

Color signalling can also be obtained through use of switched diodes in series with the LEDs so that only one color at a time is on. The diodes are switched in and out by means of switches or transistors and can be made to follow a programmed sequence or at a fixed clock rate. The new two-color red-green LED in a single epoxy case can also be used in the spinning display so that the number of switches and LED units can be reduced.

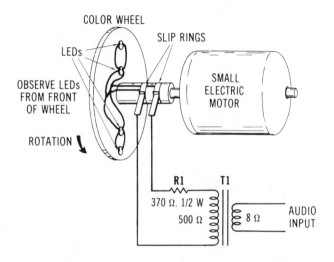

Fig. 10-5. LEDs mounted on color wheel which are powered through slip rings from transistor radio or stereo amplifier.

LED Light-Bar Module

A device introduced by Hewlett-Packard, Inc., can be used to good advantage by applying movement to it. Hewlett-Packard has a family of LED light-bar modules which are designed for use as backlighting or display panels for electronic instruments, computers, office equipment, and automobiles. The light-bar modules provide large, bright, uniform, light-emitting surfaces and they mount easily in printed-circuit cards or sockets. There are two sizes available: 0.835 × 0.15 inch (2.12 × 0.38 cm) and 0.75 × 0.15 inch (1.90 × 0.38 cm). Colors available are red, green, and yellow, and they are the largest continuously illuminated surfaces using LED technology ever offered by HP. Fig. 10-6 shows light-bar modules which can be used as illuminated legends, indicators, bar graphs, and light switches. The cost is less than $2.00 for the largest module.

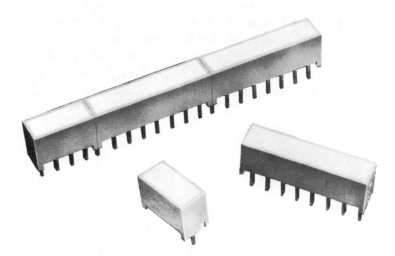

Fig. 10-6. LED light-bar modules suitable for moving displays.
Courtesy Hewlett-Packard, Inc.

Voice Recognition

The possibility of voice recognition using the SpinLED is evident because certain unique individual voice frequencies are discernible in the dynamic display. You can pick out different speaker voices from among several while you are observing the display. It may thus be possible for certain fixed spoken phrases to be passed by telephone and observed visually by a deaf person.

An inexpensive spectrum analyzer is possible using the SpinLED, as unique frequency energy in an audio signal can be resolved separately into a dot-dash pattern. What sounds like power-line noise to the ear on a broadcast-band receiver or TV receiver can be observed to follow a repetitive pattern and various pulse energies separately observed on the SpinLED.

Stereo Channel Separation

LEDs mounted on a vibrator-type device using a piece of hacksaw blade can serve as a four-channel level indicator in a high-fidelity stereo amplifier system. With this display it is possible for the listener to see what he or she hears so that he or she can make instant adjustment with left/right and front/rear level controls. In certain applications this type display can replace a cathode-ray tube with its attendant high-voltage supply and deflection driving circuits.

Frequency Measurement and Synchronization

Frequency measurement and synchronization is possible through use of two LEDs vibrating back and forth in unison, one powered from a reference source and one from a source to be measured and synchronized. The reference source will strobe out a series of dashed lines with fixed on and off periods. The frequency to be measured will strobe out a pattern so that one can tell if the frequency is below, above, or near the reference frequency. As shown in Fig. 10-7, when the measured frequency is low (Fig. 10-7A), there will be fewer long dashed lines compared with the reference frequency. When the frequency is higher (Fig. 10-7C), there will be more shorter dashed lines than the reference. As the measured frequency is adjusted to be near that of the reference frequency the number of lines strobed out by each source will be the same (Fig. 10-7B). Using the process of observing "visual beats" between the two strobed LEDs it is possible to adjust one frequency to within 0.1 Hz of the other (one "beat" in 10 seconds). The vibrating LED display, which can be a handheld or panel-mounted device the size of a panel meter, makes use of flexible wire leads to the LEDs so that slip rings are not required.

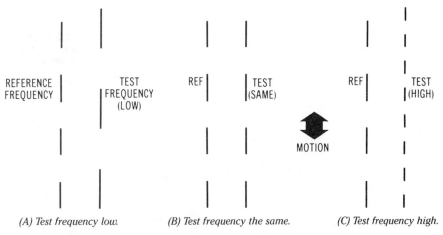

(A) Test frequency low.   *(B) Test frequency the same.*   *(C) Test frequency high.*

Fig. 10-7. Comparing test frequency with reference frequency using vibrating LEDs.

Digital 1s and 0s

The vibrating LED can be used to observe that a digital circuit component is switching states between 1s and 0s. Ordinarily this cannot be observed with an incandescent lamp or LED, although the LED which is intermittent would put out less light than an LED which is on continuously. With the vibrating LED the ratio of *on* to *off* time can be easily observed. An *Idle-Busy* time display for a computer can be built very easily without having to resort to an oscilloscope display.

Rotating Machinery

Attached to a rapidly rotating or moving part of a piece of machinery an LED can be made to "stop" by synchronizing the rotary motion with a variable-frequency voltage source. The mass of the LED is very low, which permits it to be mounted to the moving part, with the low-voltage power source supplying the LED voltage through slip rings or a rotary induction coil arrangement. The SpinLED in the form of the moving LED requires no special precaution with regard to voltage as it will operate at a nominal 1.6 to 2 volts. Because of these features a low-cost solid-state dynamic display is within the reach of the high-fidelity music listener as well as industry.

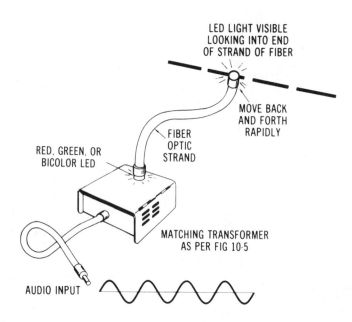

Fig. 10-8. Moving fiber optic cable or strand viewed on end appears as if the LED were moving.

Vibrating Fiber Optics

Fiber optics are light-wave guides. Instead of electrons moving through a piece of copper wire to carry information, light moving through a fiber optic cable carries the information. The cables can be used to carry the light information to the flashing LED. Instead of moving or vibrating the LED the fiber optic cable is placed against the LED and the other end of the fiber cable vibrated. Thus, when you look at the vibrating end of the cable it flashes in the same manner as if you were looking at the LED itself. Refer to Fig. 10-8. These cables can be many tens of feet long without loss of light power. Fiber optics is wide open for innovation and the vibrating optic cable is one such avenue for exploration.

11 Sound Over a Light Beam

Since the beginning of history light has been used as a means of communication. The Greeks and Romans used large bonfires to signal from hill to hill until a message was received at some distant headquarters. The presence of the light signaled the intent of the message, and not the intensity of the light: hence, "One if by land, two if by sea." Imagine what might have happened if the signal was: "One bright light if by land, one dim light if by sea." What would have happened to the message if some fog had set in and dimmed the one lantern? It is not easy to tell a bright light from a dim light without some reference source. This is one reason why light is usually not modulated in intensity for signalling to the eye. Also, we say that a light or candle "flickers" because the change in intensity occurs at a slow rate. When the intensity changes exceed 10 to 12 per second, the eye can no longer detect a change. Briefly, this is why the motion picture works. We do signal in the *frequency* domain, though, where *green* is go, *red* is stop, and *yellow* (or *amber*) is caution.

With the right equipment, however, there are a number of things we can do. In 1880 Alexander Bell used his *Photophone* to talk via a reflected light beam from the sun. He used a small flexible mirror at the end of a tube into which he spoke. The light from the sun reflected from the mirror in the direction of the receiver was thus amplitude modulated by this process.

Perhaps you have heard of demonstrations that used flashlights which were modulated by a microphone and the signal was picked up hundreds of feet away. These demonstrations that other people did, while interesting, do not serve the same purpose as when you are able to try a simple experiment yourself. Therefore in this chapter we will cover some of the basics of light-beam communications.

The Transmitter

Let's start by considering what constitutes a light transmitter. It is very simple in nature: an incandescent lamp, a candle, an oil lamp, an LED, infrared LED, laser, the sun, a fluorescent light—all of these are transmitters of light.

The Light Bulb

In Fig. 11-1 we see a block diagram for a light transmitter that is modulated in intensity (amplitude modulation) as is used in the am broadcast band. In our case the transmitter frequency is at light frequencies, which cover from 400,000,000 MHz (red) to 810,000,000 MHz (violet). The am broadcast band covers from 0.55 to 1.6 MHz. Both of these transmitters radiate electromagnetic waves. One we know as light waves, one we know as radio waves. The lamp shown in Fig. 11-1 is slow to heat and slow to cool, as you might have observed at a sports field when the huge incandescent lights are turned off at night. The 1000- to 5000-watt bulbs take 5 to 10 seconds to cool to the point where you can no longer see them glow. In the same sense the lamps (household bulbs or flashlight lamps) are slow to heat and cool so that they cannot closely follow the modulation frequencies. That is, they cannot instantly be switched on and off; there is a time lag.

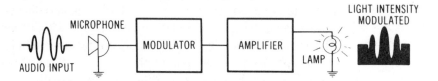

Fig. 11-1. Lamp that is intensity modulated by an audio signal.

The LED

One of the marvels of the LED is that it can be turned on and off very rapidly, on the order of a nanosecond, and thus is capable of a high rate of modulation. An LED is low-cost, lasts 100 years, and is easily adapted for use as a light transmitter.

Let's take a look at how we can easily and conveniently arrange for an LED to be modulated. In Fig. 11-2 we have replaced the microphone (Fig. 11-1) with the audio output from a transistor or table radio, and the light bulb (Fig. 11-1) with a red LED or infrared LED. We use a transistor radio to provide audio for us so that we can always provide a modulated audio signal to the LED while we make circuit adjustments to the transmitter or receiver.

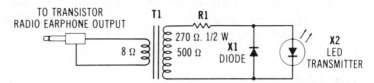

Fig. 11-2. LED transmitter driven from am receiver audio output.

The Laser

The laser is probably the fastest flash device yet devised. The energy contained in its beam of light can burn holes through thick metal. It is for this reason that safety goggles must be used when working near a laser. Since the laser is beyond the reach of most of us we will discuss only lower-powered devices.

The Modulator

We have discussed amplitude modulation of a light bulb and an LED using a microphone or a radio. There are other fairly simple means of modulating light which we will now discuss.

The Flashlight

The light from an ordinary flashlight can be modulated in a very simple manner so that it can easily be received and amplified through an ordinary amplifier. The flashlight is turned on and aimed at a solar cell (and other devices to be described) and the flashlight tapped gently near the bulb housing. A gentle *ping* will be heard each time the light is tapped; it lasts for a second or so. The tapping causes the bulb filament to vibrate so that its light output is amplitude modulated, and this effect can be heard up to 20 feet or so, using a solar cell as a detector.

Chopping Light

In addition to "pinging" a flashlight the light from the flashlight can be modulated by "chopping" its light output. This is done by passing the hand rapidly back and forth in front of the beam with the fingers extended and open. This action "chops" the light at a low audio rate, but what you are doing is amplitude modulating the output of the flashlight. This type of signal modulation can be picked up 20 to 30 feet away, depending on the size of flashlight used and whether it is dark outside. In the daytime much of this effect may be masked by the bright background from the sun. You can also use an electric fan to chop the light. For a four-bladed fan the chopping rate is four times as great as that of the fan revolutions per minute; that is usually around 100 to 200 Hz.

The LED

Light from the LED can also be chopped with the open hand or electric fan, as well as being modulated by a microphone or transistor radio as discussed earlier.

The Receiver Detector

So far we have managed to modulate light energy in some manner. We will now look at some means of demodulating the intelligence which we put on the light-frequency "carrier."

The Solar Cell

Silicon or selenium solar cells are very good for detecting light energy and converting it into electrical energy which can be amplified so that it can be heard in a small amplifier. Fig. 11-3 shows a block diagram for a solar cell connected to a small audio amplifier. A solar cell will produce about 0.5-volt dc in bright sunlight or under a bright electric lamp. When suitably amplified the solar cell output will produce room volume for most of your experiments. Solar cells are shown in Fig. 11-4.

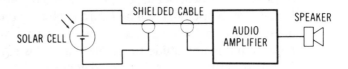

Fig. 11-3. Solar cell connected to audio amplifier to listen to light that is intensity modulated.

The Photocell

We took a close look at the photocell in Chapter 6 and Chapter 8, where we used it for light-sensitive audio oscillator and in a Wheatstone bridge. In those experiments we used the photocell for its unique characteristics— extreme sensitivity at low light levels and wide change of resistance over wide light-level changes. But in those experiments we did not test the photocell in the frequency response domain. In this area it is not very speedy in response. It was fast enough, though, to follow the buzz caused by a TV picture tube as it generates the 30 frames per second it takes to produce a picture (two rasters per frame). Because of its slow response the photocell is not ordinarily used for listening to light.

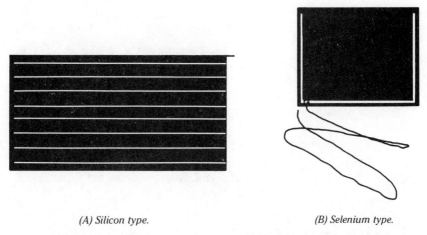

(A) Silicon type.   (B) Selenium type.

Fig. 11-4. Two types of solar cells. *Courtesy Radio Shack, Div. of Tandy Corp.*

The Photodiode

A silicon photodiode is a very sensitive detector of light and it has an extremely fast response time (less than 1 nanosecond). The photodiode can be connected directly to a miniature plug so that it can be plugged into an audio amplifier. Fig. 11-5A shows this arrangement.

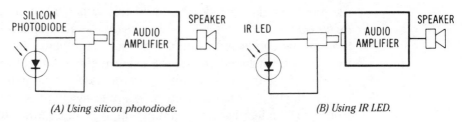

(A) Using silicon photodiode.   (B) Using IR LED.

Fig. 11-5. Detectors for intensity-modulated light.

The Phototransistor

The phototransistor is essentially the same as a regular transistor except that it is made of clear plastic, or epoxy, so that the light-sensitive surface of the semiconductor is exposed to light. The phototransistor may or may not have a wire connected to the base since light photons take the place of the lead wire. Fig. 11-6A is the schematic for a phototransistor as input to an amplifier, and Fig. 11-6B shows a phototransistor with a Darlington amplifier. All of these various light detector-sensors are available at the local neighborhood radio parts supply houses, and they are very reasonable in price.

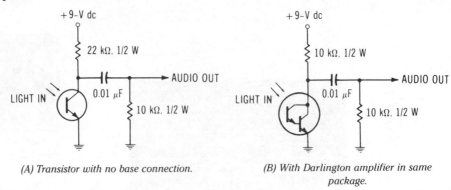

(A) Transistor with no base connection.   *(B) With Darlington amplifier in same package.*

Fig. 11-6. Phototransistors as light detectors.

The Audio Amplifier

There are a number of audio amplifiers which provide sufficient gain in order to conduct the simple tests or experiments described so far. We will consider only the following three amplifiers.

The Humbug

Of the audio amplifiers or modules available from various sources, one of the more popular was described by this writer in *Listen To Radio Energy, Light, and Sound.* This book describes in detail a number of experiments that can be conducted using a high-gain, handheld, audio amplifier called the *Humbug.* Included among the experiments relating to light is the use of a 6-inch reflecting makeup mirror to increase signal gain in order to listen to automobile headlights from a half-mile away, to aircraft landing lights from 5 to 10 miles away, to lightning up to 30 miles away, to malfunctioning mercury-vapor and neon lights which the eye cannot see but the ear can hear, to the light from a burning candle, to a camera photoflash, to light backscattered from clouds, and many experiments with light. Other experiments are described in the book which cover energy in other domains, such as the induction field, radio, sound, and the magnetic field. This book is now out of print, but you might find a copy at your local library.

The Humbug uses an amplifier made for Radio Shack and available at some stores. The amplifier has the most audio gain yet available for the average electronics buyer. Ask for Radio Shack Catalog Item No. 277-1240, four-transistor audio amplifier. Fig. 11-7 shows the Humbug amplifier, which is our first possibility.

Fig. 11-7. The Humbug high-gain audio amplifier.

One-Watt Speaker-Amplifier

This self-contained battery-operated amplifier is available from Radio Shack stores and has sufficient audio gain for most experiments. It has an audio input jack for inputs, such as from a solar cell, and it has a speaker output. It is Radio Shack Catalog Item No. 32-2031.

Portable Phone Listener

The portable listener is a transistorized amplifier that can be used with a telephone to amplify conversations. It can be used with an induction pickup or any of the other sensors described. The Radio Shack Catalog Number is 43-231. It operates off a 9-volt battery.

A Simple Light Communications System

We have learned a bit about the various parts of a light communications system. Let's now connect the three parts of the system together and make it play. The three parts we will connect are the modulated LED, the solar cell, and the audio amplifier. Through this simple system you will play some music, and this will make you feel really good, mainly because you have accomplished it yourself. In Fig. 11-8 we see a complete light communications system that we can use to transmit music or speech over a distance of several feet, depending on the amount of light from the LED and the gain of the audio amplifier. While very simple in nature this system contains all of the components of a fully working system that industry uses in communicating by means of light beams.

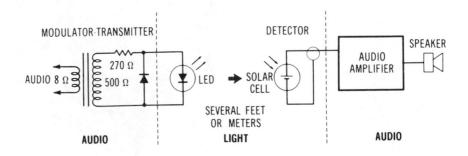

Fig. 11-8. A light communications system using an LED transmitter and solar cell receiver-amplifier.

Fiber Optics and Light

In this section we will briefly discuss the use of fiber optics and how we can replace the light communications path of free space with that of several feet of fiber optic material.

That light will follow a curved or bent path in an optical material was first demonstrated by John Tyndall in the late 1800s when he discovered that light was guided along an arc by a stream of water. Nothing much has happened in fiber optics until recent years. Since then a number of things have been happening in research where almost daily a new breakthrough is announced in telephone or video communications.

In Fig. 11-9 we see light entering the end of a fiber optic cable or strand. The strand, fiber, or filament, which can vary from a few thousandths of an inch to about a ¹/₁₆ inch in diameter, is clad or covered with a material having a refractive index such that light is reflected or bounced off the sides of the walls of the fiber. The reflection off the walls of the fiber is akin to throwing a tennis ball down a long, hollow, sewer pipe, with a number of bounces taking place off the walls until the ball emerges at the end.

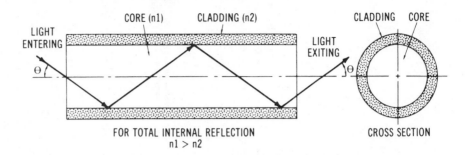

Fig. 11-9. Light entering and leaving a fiber optic strand.

The fiber optic cable or strand is made by means of a special technique where glass, fused silica (quartz, like ordinary sand), or plastic is melted so that it can be drawn out into long continuous filaments (much like taffy candy is pulled), some as long as many thousands of feet. After that, the cladding, or outer covering, is added.

The loss associated with fiber optic cable is very low compared with copper coaxial cable. Light rays which do not reflect back into the wave-guiding core are reflected into the cladding (the cladding mode) and are eventually lost by radiation out of the cladding into space. Commercially available light cable has an attenuation factor of 10 dB per 1000 meters (3280 feet) as contrasted to coaxial cable loss of 10 dB per 30 meters (100 feet) at 1000 MHz. Not only is coaxial cable very expensive, it is very lossy.

Fiber optic cables can be bought at some radio supply stores as decorator lamps with an electric lamp or flashlight at one end of the bundle and a spread-out flower display at the other end of the bundle. A short length of one of these cables can be used to transmit light for demonstration purposes. In Fig. 11-10 is shown a short length of fiber placed between the transmitting LED and a solar cell or phototransistor. Fig. 11-11 shows a matched pair of an IR LED and a phototransistor.

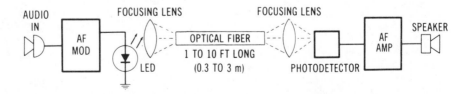

Fig. 11-10. Communications by means of a fiber optic cable.

Fig. 11-11. Matched pair of IR LED emitter (left) and phototransistor detector (right).
Courtesy Radio Shack, Div. of Tandy Corp.

You will be quite pleased with yourself once you have conducted this simple experiment. This easy approach will greatly assist you in understanding some of the basics of LEDs, fiber optics, and light detectors. Lenses at both ends of the link will increase the signal-to-noise ratio so you can carry the light farther, in a longer piece of cable. May you successfully see the light at the end of the fiber optic tube!

12 A Wide-Resistance-Range Audio Continuity Tester

This audio continuity tester, which uses a low-current circuit and has a wide resistance range of test, is designed around a property of the familiar 555 timer IC, viz, it will function as a multivibrator over a wide frequency range if the value of one resistor is changed. Conversely, it can be made to provide an audio output over a wide range of resistance values—from 0 ohms to over 30 megohms.

Operation

Fig. 12-1 shows the test leads to the audio continuity tester shorted together. The "short," or 0 ohms, produces an audio tone from the self-contained speaker of about 7 kHz. Actually the audio is 7000 pulses per second (pps), but the steady stream of pulses sounds like a tone (sine wave).

When we now place a resistor of high value at the test probes (a nominal 100 kilohms to 1 megohm) the audio tone drops rapidly to a fast putt, putt, putt. When we place a 20-megohm to 30-megohm resistor (two or three 10-megohm resistors in series) across the test leads we will get a one putt-putt per second output. To show that you are actually sensing the 30 megohms remove the test probes so that you have an "open" (infinite resistance) and you'll see that the oscillator stops oscillating. It could be that the 555 timer oscillator audio tester you build might actually be able to tell the difference between 50 megohms and an open, that is, "to putt or not to putt."

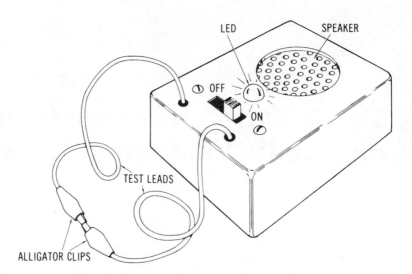

Fig. 12-1. Audio continuity checker with leads shorted.

Let Your Ears Do the Walking

In Chapter 9 we discussed the operational usefulness of an audio continuity and voltage tester. The audible continuity tester is most worthwhile for testing devices without having to glance from a test probe location (a light bulb) to an ohmmeter and back to the next probe point. It is most helpful in a test mode when a *go/no-go* test condition is sufficient, that is, when we need to know only that a device is good or bad and we are not trying to *measure* the resistance value of the device. So we let our ears do the walking from the test point to the sound; we do not have to move our eyes from the test point.

Testing Bidirectional Devices

The audio continuity tester is most valuable for testing bidirectional devices such as light bulbs, lamps, resistors, transformers and windings, fuses, and circuit wiring. These devices permit current to travel in both directions without change, whereas a diode (or other unidirectional device) does not conduct current equally well in both directions (because of the front-to-back resistance ratio) and a difference in audio tone level will be noted. The audio tester is useful for testing low-current devices, and it will indicate forward and reverse continuity conditions for diodes, transistors, capacitors, LEDs, and the like.

Circuit Analysis

The circuit diagram for the audible resistance tester is shown in Fig. 12-2. Note the similarity of the circuit to that of Chapter 6, the light-sensitive audio oscillator. In the latter circuit we used light to set a value of resistance of a photocell and that set the audio frequency of the 555 timer.

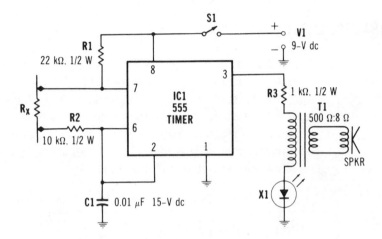

Fig. 12-2. Circuit diagram for the audible continuity tester.

The Unknown Resistance R_x

In the present circuit we use the value of an unknown resistance, R_x, to set the frequency of oscillation. From the tone, high or low, we determine whether the resistance of the device we are testing is low or high. The schematic diagram shows where R_x is connected into the multivibrator circuit. Resistance R_x can be any value from 0 ohms to over 30 megohms.

Low Current in Test Probes

Very little current flows in the test probes so that when R_x is zero ohms the current through the probes is about 270 microamperes. When R_x is 1 megohm the current through the probes is about 9 microamperes.

555 Operates in Astable Mode

The 555 timer is operated in the astable oscillator mode, where the free-running frequency and duty cycle are both accurately controlled with three external resistors and one capacitor. The external capacitor, C1, charges through R1, R2, and R_x, and discharges through R2 and R_x only. Resistor R2 is used to limit the upper frequency of oscillation to about 7000 (7 kpps) pps when R_x is 0 ohms; otherwise the frequency would be out of the upper range of hearing (beyond 18,000 pps). The low-frequency range of approximately 1 pps is set by the value of R_x when it is above 20 megohms or so. An open circuit produces no output as the input resistance, R_x, is infinite. The total current drawn from the battery is about 7 mA.

Construction

The continuity tester can be mounted in any plastic case with the speaker mounted in the top center. Refer to Fig. 12-1 for a layout of the LED pilot light, the battery on/off switch, and the holes for the two test leads. Tie a knot in the test leads where they enter the case so that they will not pull out or strain the electronic components. Table 12-1 is the parts list for the audio continuity tester. The components are almost identical with that of the light-sensitive audio oscillator (Table 6-2).

Table 12-1. Parts List for Audio Continuity Tester

Item	Description
V1	9-V Transistor Radio Battery
C1	Capacitor, 0.01-μF, 15-V dc
C1	555 Timer IC Chip
X1	Red or Green LED
R1	Resistor, 22-kΩ, $\frac{1}{2}$-W
R2	Resistor, 10-kΩ, $\frac{1}{2}$-W
R3	Resistor, 1-kΩ, $\frac{1}{2}$-W
T1	500-Ω : 8-Ω Transformer
TL	Test Leads, 24 to 36 inches (60 to 90 cm) with Alligator Clips
SPKR	Speaker, 8-Ω Miniature
S1	SPST Switch, On/Off
Misc.	Plastic Utility Case, 9-V Battery Connector Clip, Hookup Wire, Solder, 16-Pin IC Chip, PC Card, etc.

Testing for Operation

After you have completed the electrical wiring, check it against the schematic to see that all connections have been made. If all looks good, turn the on/off switch off and insert a 9-volt battery in the unit. Turn the unit on and touch the two clips together. If the circuit is wired correctly the speaker will issue a tone of about 7 kpps and the LED will be lit brightly. When the probes are held open there should be no sound coming from the speaker and the LED may be on or off, depending on where the multivibrator flip-flop flopped: on or off. As you try different resistor sizes the frequency of the tone should decrease as the resistance under test increases. Look through your junk box and hook some 1- to 10-megohm resistors and try them to see how high a resistance you can test and still have the oscillator flip states slowly. For a very high resistance (30 megohms) the LED will flash on and off very slowly and you should hear a click each time the LED flashes on *or* off. If you look at the speaker cone you can see it pull in and out as current flows through the speaker winding. Approximately 3 to 4 mA of current flows through the speaker winding and LED.

Additional Uses

A number of uses have been outlined for the audio continuity tester. We can now examine additional uses for the audio tester and see if we can't find other uses where an audio tone will alert us to some specific event. Some of these uses are outlined below but it won't be long before you think of additional ones yourself. Once you have aroused your curiosity, you won't be satisfied until you have done it yourself. Good luck!

Water Seepage in Boat or Home

Water or gasoline in a place where it shouldn't be may be cause for alarm. You can sense this by using two nails driven into a piece of varnished wood, or two copper lines etched in a pc card, or any insulated two-element device that you can place in the basement as the sensor to let you know things are dry and alright. Refer to Fig. 12-3 for details of the sensor. When the two nails become immersed in water because it is rising in the basement, the sounder will sound out because the resistance across the two nails goes from a very high value to a somewhat lower value. Impure water or gasoline is a good conductor of electricity and the continuity tester will sound out to alert you. And of course, line lead length between the sensor and the audio continuity tester is not important for operation; you can run thousands of feet and still tell the difference between an open and a short.

Liquid-Level Height Detector for the Blind

A liquid-level detector can be easily made out of two pieces of stainless-steel wire or other material that is hung over the edge of a cup or glass. When

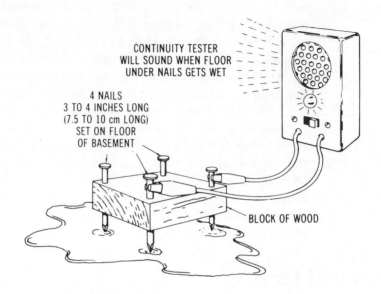

Fig. 12-3. Water or gasoline sensor for boat or house.

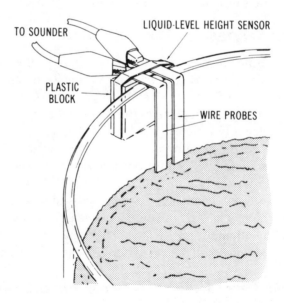

Fig. 12-4. Liquid-level height detector sensor will cause sounder to signal when level of liquid reaches it.

liquid is poured into the glass or cup the sounder will emit a tone as the two elements of the sensor are immersed in the liquid, be it water, coffee, tea, or hot soup. Fig. 12-4 shows the details of the sensor. A design similar to this can be used to tell you when a pot of water is boiling because the liquid level will boil or bubble across the sensor elements and the sounder will sound intermittently as the liquid level rises and falls rapidly.

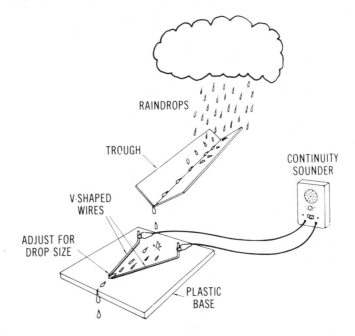

Fig. 12-5. Raindrop detector will emit tone when water drop reaches V-shaped wire sensor.

Raindrop Detector

You can detect raindrops and rain by placing the two elements of the sensor made out of copper wire in such a shape that a drop of water covers both the elements. The sounder will sound out each time a drop of water rolls over the probes, and it will continue to sound out until the drop rolls off the probes or evaporates. Refer to Fig. 12-5 for details of the sensor.

13 A Light-Sensitive Detection Device: The Lightstalker

The Lightstalker is an easy-to-build device that can be placed in a window to alert you if a would-be intruder is prowling around outside your home. The device uses passive light from a nearby or distant source such as a street light, a neighbor's porch light, or some other exterior light such as a playground, tennis court, shopping mall, or auto freeway. If someone walks between the light source and the *Lightstalker*, the incoming light intensity drops, which, in turn, sets off an audible alarm.

The Lightstalker uses an unusual comparator circuit to make it more sensitive than most inexpensive light detectors. It is designed to work in dimly lit areas where the light source may be as far away as one hundred feet or more.

Operation

The Lightstalker and its power transformer are shown in Fig. 13-1. Mounted on the metal backplate of the Lightstalker are a switch, a light emitting diode (LED) and a piezosounder. The switch is used to select the LED or the piezosounder as the output indicator when the unit has been tripped. The piezosounder is the desired output indicator under normal operation. The LED, on the other hand, is useful when first setting up or experimenting with the Lightstalker because the sounder will emit a signal until it adjusts itself to the available light level. On the front side of

Fig13-1. Back of the Lightstalker showing LED, alarm switch, piezosounder, and wall-mounted transformer.

the Lightstalker, a cadmium sulfide (CdS) photocell is mounted near the top of the case to point in the general direction of the light source.

Initial Setup

Operation of the Lightstalker is very simple. After dark, it should be placed in a window with the photocell oriented in the direction of a light source. The transformer (Fig. 13-1) should then be plugged into a ll5-V ac wall receptacle. The selected output indicator (LED or piezosounder) will then be activated for a minute or more as the Lightstalker adjusts its very sensitive threshold voltage to the ambient light intensity striking the photocell. Because of the one or two-minute adjustment time, you may find the LED to be the preferred output indicator when first applying power to the Light-stalker as the sounder output is very loud. After the unit has adjusted to the ambient light level, it is ready for operation. The LED will go out and you can then switch to the sounder. The LED will remain off when the sounder is switched into the circuit.

Ideally, the Lightstalker should be used in a dimly lit area where only one light source is present and where the light source is approximately one- to two-hundred feet away. Actually, that rarely will be the case. You may choose to use it in an area where there are several light sources or in an area that is not very dark. You can improve the Lightstalker's performance in these areas by installing a short one or two-inch piece of dark tubing (plastic, cardboard, or rubber) over the photocell to act as a telescope and block out unwanted light. If the Lightstalker still seems to be desensitized by too much ambient light, try covering the end of the tubing with a piece of black tape so that only a small hole (perhaps ⅛-inch in diameter) allows light to strike the photocell.

Color Sensitivity

Depending on the intensity of the light and the type of light in use, the Lightstalker will often work even when the light source is as far away as 500 feet. Incandescent lights seem to be the best since their spectrum approaches that of the sun's light, containing red, green, and blue energy. Mercury-vapor lights, which are normally very bright, work well if they are no more than one- to two-hundred feet away. Mercury-vapor light is in the blue spectrum and we believe the Lightstalker's sensitivity to mercury-vapor lights is reduced by the photocell's response curve. Cadmium sulfide (CdS) photocells are most sensitive to green, yellow, and orange light (from an incandescent lamp) and less sensitive to blue (from a mercury-vapor light).

Circuit Analysis

Let's take a look at how this unique passive intruder detector works.

Power Supply

The circuit diagram for the Lightstalker is shown in Fig. 13-2. The power supply consists of wall-mounted transformer T1, fuse F1, bridge rectifier CR1, and capacitor C1. T1 is a plug-in transformer similar to the types used for games and toys. It provides 24-V ac to the 4-diode bridge rectifier, CR1. After CR1 rectifies the voltage, C1 filters the 120-Hz ripple to provide +28-V dc to the remainder of the Lightstalker circuitry.

The remainder of the circuitry can be separated into four sections: a photocell voltage divider (R1, R2, and R3), an amplifier/comparator (IC1), a threshold storage/feedback circuit (R4, R5, R6, C2 and C3), and an output circuit (CR2, S1, PS1, X1, R7, R8, and R9).

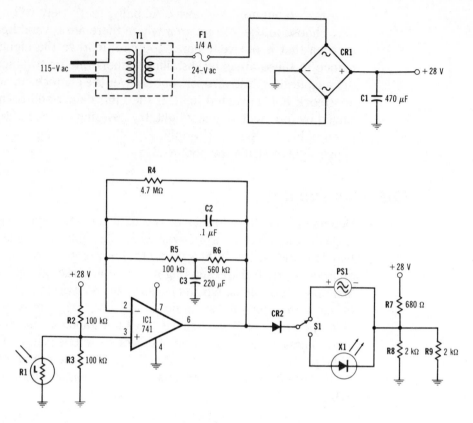

Fig. 13-2. Circuit diagram for the Lightstalker. The power supply is at the top and the main unit at the bottom.

Output Circuit

Looking at the output circuitry first, notice that switch S1 is used to select the piezosounder (PS1) or the LED (X1) as the output device. Resistors R7, R8, and R9 form a voltage divider network connected to one end of PS1 and X1. The voltage at the junction of R7, R8, and R9 is +17 volts when no current is flowing through PS1 or X1. When the output of IC1 exceeds +17 volts, the piezosounder or LED is activated because current will then flow from the junction of R7, R8, and R9, through PS1 or X1, and through S1 and CR2.

The Photocell

Let's look now at the voltage divider which is formed by photocell R1 and resistors R2 and R3. The photocell voltage divider provides a voltage to the noninverted (+) input to IC1 that is proportional to the incoming light intensity. When the Lightstalker is exposed to bright light, the photocell resistance is low, and the voltage at IC1's noninverted (+) input is also low (a few volts or less). When exposed to dim light, the photocell resistance is high, and the voltage at the noninverted input to IC1 is also high (close to +14 volts).

The Comparator

IC1, the amplifier/comparator, compares the voltage from the photocell with a threshold voltage coming from capacitor C3 and resistor R5. If the voltage from the photocell is more positive than the threshold voltage, the output from IC1 (pin 6) will increase significantly, to a level of about +25 volts (which will activate the piezosounder or LED). Whenever the voltage across the sounder is positive with respect to the negative terminal, it will sound (or the LED will light). If the voltage from the photocell is less than the threshold voltage, the output from IC1 will decrease significantly (approaching zero volts). If the two input voltages are about equal, the output from IC1 will remain at a level approximately equal to the level of the two inputs.

Threshold Circuit

We will now look at the threshold/feedback circuit. As previously mentioned, the threshold voltage that is applied to the inverted (−) input of IC1 comes from resistor R5 and capacitor C3. This capacitor is charged to a voltage that is equal to the ambient light level from the light source. The charge on C3 is increased or decreased by the current that flows through resistor R6.

When power is first applied to the Lightstalker, the threshold voltage on C3 will be zero volts. Let's assume that the ambient light from the light source results in a level of +10 volts from the photocell voltage divider. Clearly, the voltage from the photocell is more positive than the threshold voltage, so the output from IC1 increases to +25 volts. The output voltage of +25 volts activates the output device, but more importantly, it allows current to flow through R6. The current flowing through R6 causes C3 to charge slowly, which, in turn, makes the threshold voltage increase slowly.

The threshold voltage continues to increase as C3 charges. Because of the RC time constant for R6 and C3, it can take a minute or more for the threshold voltage to reach the photocell voltage. Once the threshold voltage equals the photocell voltage (+10 volts in this example), the output from IC1 drops to a level equal to the two input voltages. C3 then stops charging and the threshold voltage remains at a level equal to the photocell voltage.

If someone now walks in front of the Lightstalker (breaking the light beam), the photocell voltage will increase, causing the output from IC1 to increase, which, in turn, will activate the piezosounder or LED.

Flicker Filter

Two components in the threshold storage/feedback circuit have not yet been discussed: resistor R4 and capacitor C2. Resistor R4 provides negative feedback from the output of IC1 to its inverted input. Without R4, the Lightstalker has a tendency to oscillate at about 1 Hz.

To understand the function of C2, we must first consider the light sources used by the Lightstalker. The lights are powered by 60-Hz power, and while we cannot see it, they flicker at a 120-Hz rate (once for each half of the 60-Hz sine wave). The function of capacitor C2 is to provide negative feedback so that the 120-Hz signal will not be amplified by IC1. Without C2, a light source with a strong 120-Hz signal can sometimes cause the LED or piezosounder to be activated all the time.

Construction

The parts needed to construct the Lightstalker are shown in Table 13-1. The Lightstalker is built in a plastic case with the switch, LED, and piezosounder mounted on the backplate as shown in Fig. 13-1. The photocell is mounted on the front of the unit, near the top to face the light source.

Method of Construction

We used a general purpose "Experimenter's Dual IC Board" to mount the integrated circuit, capacitors, bridge rectifier, diode, resistors, and fuse with holder. Component placement is not critical; however, you must be sure that the larger components on the circuit board are not positioned in such a manner as to obstruct installation of the backplate and piezosounder.

Power Transformer

A plug-in wall-type transformer is used to power the Lightstalker. The circuit requires between 20- and 24-volts ac at less than 100-mA current. To prevent overheating, we selected a transformer that can provide up to 510-mA at 20-volts ac. Eight feet of common speaker wire were used to connect the plug-in transformer to the Lightstalker circuit. A small hole is drilled on one side of the plastic case to allow the wire to pass through the chassis.

Table 13-1. Parts List for Lightstalker

Item	Description
C1	Capacitor, 470-μF, 35-V dc
C2	Capacitor, 0.1-μF, 50-V dc
C3	Capacitor, 220-μF, 35-V dc
CR1	Bridge Rectifier, 1.4-A, 100-piv
CR2	Diode, 1N4001 or Equivalent
F1	Fuse, $^1/_4$-A, 250-V
IC1	Integrated Circuit, Type 741 Op Amp
PS1	Piezosounder, Radio Shack 273-068 or Equivalent
R1	Photocell, Cadmium Sulfide (CdS)
R2, R3, R5	Resistor, 100-kΩ, $^1/_2$-W
R4	Resistor, 4.7-MΩ, $^1/_2$-W
R6	Resistor, 560-kΩ, $^1/_2$-W
R7	Resistor, 680-Ω, $^1/_2$-W
R8, R9	Resistor, 2-kΩ, $^1/_2$-W
S1	Switch, SPDT, Mini Toggle
T1	Transformer, Input 115-V ac, 60-Hz, 14-W; Output 20-V ac, 0.51-A
X1	Red LED
Case	Radio Shack 270-233 or Equivalent
PC Board	Radio Shack 276-159 or Equivalent
Misc.	Fuseholder, Speaker Wire, and Hookup Wire

Capacitor Quality

One point to consider when building the Lightstalker concerns internal leakage current in normal electrolytic capacitors. This leakage is normal in these types of capacitors and we must be certain that the leakage current in capacitor C3 (220 μF) is kept low because the charging current from resistor R6 is very low. If the leakage current of capacitor C3 is too high, the threshold voltage will never reach a level equal to the photocell voltage. This effect will cause the LED or piezosounder to be activated all the time. To prevent such a problem, don't use an old capacitor from your "junk box." Use a new one.

Uses for the Lightstalker

The Lightstalker is a very versatile and sensitive photosensor. Its uses are limited only by your imagination and your environment. As previously mentioned, it can be used to detect a potential thief moving around at night outside your home. It can also be used to alert you if someone enters your driveway or sidewalk after dark.

Versatile Light-Change Detector

Some modifications can be made to the Lightstalker to make it even more versatile. For example, the photocell can be installed in a second enclosure which can be mounted 50 to 100 feet from the Lightstalker. Common speaker wire can then be used to connect the photocell to the Lightstalker circuit. The Lightstalker can be placed by your bedside and the light sensor placed where it guards the approach to your home.

Another possible modification is to connect a low-current relay to the output (in place of the piezosounder or LED) and allow the Lightstalker to turn on a light or siren if activated. Some additional circuitry would be required to keep the light or siren operating for a specified period of time. Ordinarily, you will hear the sounder only while the beam is being broken.

The Lightstalker is designed to work best in dimly lit areas, but, as previously mentioned, its performance can be improved in more lighted areas by placing a cylindrical tube over the photocell. If lighting conditions are acceptable, such an arrangement could be used to alert you if someone enters a restricted area inside a building.

Add a Telescope

A small, inexpensive telescope or binocular piece can be placed in front of the photocell so that the cell sees only a small illuminated area such as a distant road or driveway. When an auto or person breaks the beam, the sounder will sound out. The greater the light intensity of the beam that is broken, the longer and louder the sounder will emit a tone as the unit readjusts itself to the changed light level.

Driveway Sensor

The Lightstalker can be aimed down a dark driveway, and when a car with its headlights on drives up to your house, the sounder will sound because the headlight beams will eventually be broken by trees, structures, or the headlights being turned off.

The Sun and Moon

A novel feature of the Lightstalker is that it is not affected by slowly changing great light sources such as the rising or setting sun or moon. This means the unit automatically adjusts itself to the rising sun if you should be looking east when the sun comes up. Likewise, the moon will not affect operation as it passes through the beam of the photocell sensor. It is only when the beam is affected (interrupted) rather quickly, such as a walking person or large animal, that it is triggered off.

Fast Moving Objects

The filtering action of resistor R4 and capacitor C2 prevents fast moving objects, such as cars or trucks, from setting off the system. Since the circuit is designed to not be sensitive to 60- or 120-Hz power line flicker, it will not see cars or trucks interrupting the light beam. However, it will see pedestrians, large dogs, and bicycle riders moving through the beam.

Extended Uses

After you have become used to the capabilities of the Lightstalker, you will think of other locations or adaptations where it can be used to serve your particular use. You'll be able to use the Lightstalker when camping out in your travel trailer or recreational vehicle where you have access to 120-volts ac and a steady, all-night light source. In the daytime, you can use the Lightstalker to observe shadows that are cast by someone walking along a sidewalk if the shadow reduces the light coming into the sensor that has been placed in the window.

14 A Visual Telephone Ringer

We are used to listening to the phone "ring" with our ears. But the phone can be made to "ring" to our eyes, also. This feature is especially important in a high noise environment. If we don't hear the phone ring, we will miss a call. But, if we can be alerted to the ring by a visual means, we can answer the phone without missing the call.

"See" the Phone When It Rings

There are times when we might not be able to hear the phone ring due to the high ambient noise level in a room, office, or shop. In the home, the TV volume could be turned up too high so that no one hears the phone ring. The vacuum cleaner could be turned on which can mask out the loudest telephone ring in the house. Perhaps the volume on a stereo amplifier is turned up and no one can hear the phone ring.

There is also the case where someone in the house might be hard of hearing and the extra noise from the TV set or radio will mask the ringing of the phone. How will that person know the phone is ringing? Then, there is the occasion when you are outside in the back yard cutting the grass and you don't want to miss an important call that you are expecting. What to do? These instances are precisely the times that you need something to alert you to the fact that the phone is ringing. You need "the visual telephone ringer" to let you know that the phone is ringing. You'll "see" the phone ring rather than "hear" it, and respond accordingly.

Operation

The visual telephone ringer makes use of a flashing 120-volt ac incandescent lamp that turns on every time the telephone rings. This means the lamp will flash on for two seconds when the phone rings and it will remain off for four seconds between rings. The ringing and lamp flashing sequence is shown in Fig. 14-1. When the telephone is answered, the lamp will stop flashing and remain off. If the phone is not answered, the lamp will flash each time the phone rings and stay off when the caller hangs up.

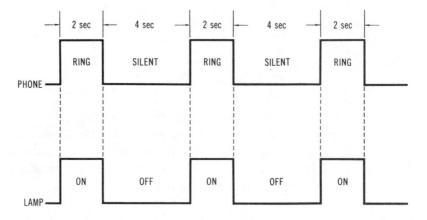

Fig. 14-1. Phone ring and lamp turn-on sequence. The lamp is on only when the phone rings.

The phone flasher is very handy both day and night. Use a lamp that is not used regularly for lighting because it will have to be devoted to being mainly a phone flasher. The lamp is left switched on but it is triggered only when the phone rings. Even in a room brightly lighted by the sun, the flashing lamp will get your attention even though you are busy doing something, such as using the vacuum cleaner, and you are not looking directly at the lamp. When the lamp flashes the first or second time, you'll cut the cleaner off and answer the phone. You might want to have several phone flashers, one for each room in which you might work where there would be a high noise level from vacuum cleaners, stereo equipment, machinery, or the like. The phone flasher is also very handy for occasions when you might want to keep the noise level in a house or office very low and yet be alerted to the fact that the phone is ringing. Such an occasion might be when a child is asleep and you don't want them to be awakened by any sounds. Your phone flasher will come to the rescue because you can have your silence and "see" the ringing, too.

Circuit Analysis

Let's take a look at the circuit that makes the visual telephone ringer (VTR) work off the phone line and the ac power line. Fig. 14-2 shows the schematic of the VTR. You will notice that the circuit makes use of an optical coupler as was used in the Sonabell, described in Chapter 4. You'll remember that the optical coupler is used to activate the circuit when the telephone is ringing. Further details on optical couplers are given in Figs. 4-2, 4-4, 4-5, and Chart 4-1. When the telephone has the ring signal on it, the optical coupler is switched on and it, in turn, switches on triac T1. When the triac is switched on, 120-V ac is supplied to the incandescent lamp and it turns on brightly.

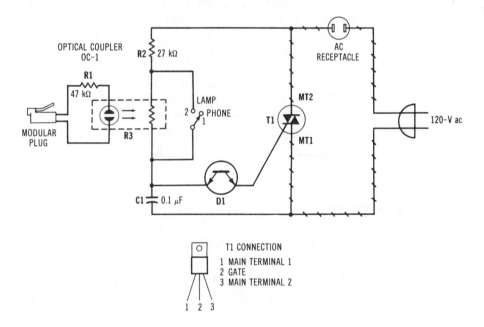

Fig. 14-2. Schematic for the visual telephone ringer.

The VTR is connected to the telephone line by means of a modular plug so that it can be easily connected to any modular jack in your home, shop, or office. Referring to Fig. 14-2, we see that resistor R1 (47 kΩ) and the neon bulb are connected across the line. When the phone is not ringing, and it is "on-hook," there is a voltage of 48-V dc across the line. Since it takes about 65-V dc to ionize the neon bulb, the on-hook voltage is not sufficient to turn it on. However, when the ringing signal voltage comes on the line from the telephone central office, it swings plus and minus ±90 volts (90-volts rms).

When the voltage swings positive, the neon bulb will ionize and turn on. When the bulb lights, the resistance of the closely coupled photoresistor (R2) drops greatly because its resistance decreases with an increase in light. A voltage appears across diac D1. This voltage causes the gate (G) of the triac to go positive and it turns on. When the triac turns on, the resistance from terminals MT1 to MT2 drops to near zero.

The triac is now on and this completes the circuit to the incandescent lamp which is plugged into the ac receptacle. The 120-V ac is applied to the lamp, causing it to light to full brilliance while the phone is ringing. The lamp has been turned on by the action of the triac, just as if you had turned on a wall switch to the lamp. The triac will turn on for each half cycle (plus and minus) of the 60-Hz power line voltage because its gate is held positive by the diac and photoresistor-neon bulb combination. Remember, the ringing voltage is 90-V rms at a 20-Hz rate while the power line voltage is 60-Hz—three times as fast. When the phone rings, the lamp will come on and appear to blink very rapidly.

To use the lamp as a lamp instead of a VTR, a single-pole, single-throw (spst) switch is wired across the photocell to short it out and provide voltage to gate G, keeping the triac turned on all the time. This switch will not affect the telephone operation in any manner; it should be labeled "Phone" and "Lamp" so that you can glance at the labels and know what the VTR is set to do.

Construction

The parts list for the VTR is given in Table 14-1. These parts are readily available at most of the local electronic parts houses. Fig. 14-3 shows the VTR mounted in a small plastic cabinet with the ac receptacle for the lamp mounted on the top. The "phone-lamp" switch is mounted on the top of the cabinet for easy access to switch the VTR from lighting on phone-ring or for full-time use of the lamp. To make a hole in the cabinet large enough to mount the ac receptacle, draw a circle on the top of the cabinet large enough to accept the receptacle. Next, mark and center-punch a number of locations around the circle so that you can drill a series of small holes through the top. You can then break out the plastic of the circle and clean up the rough hole with a file or coarse sandpaper.

The components that are not mounted on the top of the plastic cabinet are mounted on a piece of perforated board that is cut to size to fit inside the cabinet. Remember, as you connect the circuit, be sure to mark on the circuit schematic that portion of the circuit which you have just wired. This will help save time and avoid confusion because you can stop work at any

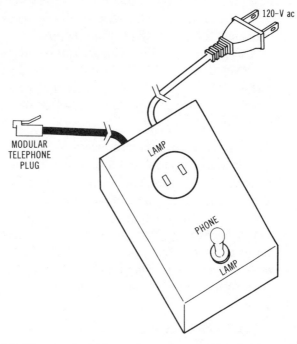

Fig. 14-3. The visual telephone ringer mounted in a small plastic cabinet.

Table 14-1. Visual Telephone Ringer Parts List

Item	Description
C1	0.1-μF, 100-V dc Capacitor
D1	1N5758 Diac, or Equivalent
NE2	Neon Bulb, NE-2 (or use OC1)
OC1	Optical Coupler, Sigma 301T1-120A1 (If not used, see NE2 and R3)
R1	Resistor, 47-kΩ, $^{1}/_{2}$-W, 20%
R2	Resistor, 27-kΩ, $^{1}/_{2}$-W, 20%
R3	Photocell, Cadmium Sulfide (CdS), Radio Shack 276-116, or Equivalent (or Use OC1)
SW1	Switch, SPST, Radio Shack 275-401, or Equivalent
T1	Triac, 200-V, 6-A, Radio Shack 276-1000, or Equivalent
Misc.	Modular Telephone Plug with Wire, Plastic Case, AC Receptacle, Line Cord with Plug

time and then resume work without having to retrace the circuit wiring you have already completed. Notice in Fig. 14-2 part of the circuit is marked with cross-hatched lines. This part of the circuit carries the current for the lamp when the triac is switched on. Since the triac and lamp receptacle can carry up to 300 watts, heavier wire should be used for this portion of the circuit. Short lengths of No. 18 or larger copper wire (stranded or solid) is sufficient for this purpose. A 75-watt bulb will be adequate for most indoor purposes. However, for an outdoors protected lamp, you might want to use a 150- to 300-watt bulb to catch your attention. Thus, if you are out mowing the lawn or in a noisy workshop, you will notice the lamp flashing on and off out of the corner of your eye.

The 120-V ac line cord available at most electronics or hardware parts houses is about 5 to 6 feet long and will readily reach from a table to the ac receptacle. The modular telephone plug and cord should be 6- to 10-feet long so that it can be plugged into the modular wall jack. Use a duplex adapter so you can plug the VTR and telephone instrument into the same jack if necessary. Wrap black plastic electrician's tape around the exposed 120-V ac connections on triac T1, capacitor C1, resistor R2, and the lamp receptacle. This precaution during assembly will prevent any leads from shorting to the 120-V ac line when all parts are assembled into the small cabinet.

Drill two 3/16-inch holes in the top side of the cabinet, one for the ac-cord and one for the telephone cord. Tie a knot in each cord where they enter the cabinet so that when you pull on the cords, the knots will provide strain relief against the cabinet.

Connection to the Telephone Line

Before connecting the modular telephone plug to the telephone line, check all your wiring once again. Be sure that all connections have been properly made and that you marked off the completed wiring on the schematic as you went along. Plug the ac cord into a wall receptacle and the telephone plug into a modular phone jack. Next, plug a lamp into the VTR receptacle and switch it on. Remember, however, that the lamp will remain off until switched on by the ringing of the telephone or switching SW1 to the "lamp" position.

Testing for Operation

The next thing to do is to call a friend on the phone and explain to him what you are doing and have him call you back. Explain that you'll let the phone ring a number of times before answering. When your friend calls, the lamp should light each time the phone rings and go out between rings. Let the phone ring a number of times and watch the lamp light each ring. It should flicker a bit each time it lights because it is being turned on and off 20 times each second. Of course, the 60 cycle-per-second flicker due to the 120-V ac cannot be seen by the eye because it is too fast. Actually, the lamp is being turned on and off 120 times per second by the 60-Hz 120-V ac. After you are satisfied that the lamp is flashing properly with each ring, pick up the phone and the lamp should go out. It should also remain out when you hang up the phone.

To test the VTR as a lamp, switch SW1 from "phone" to "lamp." The lamp should light and stay on continuously until you switch the lamp off or switch the VTR back to "phone." While in the lamp switch position, the ringing of the phone should not affect the operation of the lamp in any manner. This completes the checkout and operation of the visual telephone ringer (VTR).

15 The Neon Telelite

The neon *Telelite* is a visual flash indicator to tell you that your telephone is ringing. It is used when you are fairly close to the telephone, but do not want the ring to disturb you or cause a loud noise.

Background of the Telelite

The concept of the Telelite has been around for over 50 years; it was originally used in radio station broadcast booths where silence was essential and yet a telephone was necessary. The Telelite uses a small neon lamp that is placed across the telephone line and flashes each time that the phone line rings, very much as the visual telephone ringer. The Telelite installed in the plastic top of a shaving cream can is shown in Fig. 15-1.

Fig. 15-1. The Telelite shown installed in a plastic top from a shaving cream can.

Circuit Analysis

The circuit diagram for the Telelite is shown in Fig. 15-2. It is a very simple circuit consisting of a neon bulb and a current limiting (voltage dropping) resistor to keep the voltage applied to the neon bulb to approximately 65 to 70 volts dc. Actually, the voltage across the neon will remain fixed at the 65 to 70 volt level while the current through the bulb will vary depending on the value of resistor R1. In Fig. 15-2A we see a neon bulb with a rating of $^1/_{25}$ watt so a 220-kΩ series current limiting resistor is used at R1. A $^1/_4$-watt neon bulb is used in Fig. 15-2B so the current limiting resistor value is lowered to 22 kΩ. The NE-2H is a high brightness neon bulb which really throws out the light.

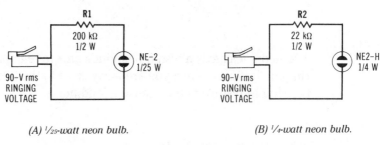

(A) $^1/_{25}$-watt neon bulb.   (B) $^1/_4$-watt neon bulb.

Fig. 15-2. Neon bulb and series resistor form a simple Telelite.

The parts list for the Telelite is given in Table 15-1. These parts are readily available at the local radio supply houses. The Telelite is especially handy for the folks that like to play the stereo real loud and yet would like to know when the phone is ringing without using a larger device such as the VTR. The unit will neatly fit in with the stereo equipment so that it can be easily seen when it flashes.

Table 15-1. Parts List for Telelite

Power	Item	Description
$^1/_{25}$-W	NE2	Neon Bulb, Radio Shack 272-1101, or Equivalent
	R1	Resistor, 220-kΩ, $^1/_2$-W
$^1/_4$-W	NE2H	Neon Bulb, Radio Shack 272-1102, or Equivalent
	R2	Resistor, 22-kΩ, $^1/_2$-W
Both	Misc.	12-ft. Modular to Spade Connections, Radio Shack 279-310, or Equivalent, Plastic Mounting Cap, Hookup Wire, Solder, Tape

A Telelite which makes use of a commercial night light is shown in Fig. 15-3. These night lights are found in numerous grocery, drug, convenience, and discount stores. They come in all sizes, shapes, and figures and are designed to plug directly into any l20-V ac outlet to provide a bright light at night. They cost less than a penny a day to operate so can be left on all the time. This unit is modified by connecting its ac plug terminal to two wires connected to a modular telephone plug, which is plugged into the telephone line. The neon light then operates off the 90-volt rms ringing voltage when the signal comes in. When there is no ringing voltage, the on-hook voltage (48-volt dc), or the off-hook talk voltage (6-volt dc), is not enough to ionize the neon bulb. Thus, it remains off except when the phone rings. When the bulb is not ionized, the circuit appears as an open and will not affect the telephone line in any manner.

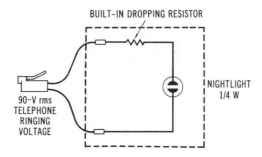

Fig. 15-3. Neon night light used as a Telelite.

A Loudspeaker Signal

Another use for the Telelite is in an application where there might be five to ten radio receivers or circuits in operation at different times. Aircraft control towers and marine shore-to-ship radio stations use a neon bulb across the loudspeaker to indicate which radio receiver is active and receiving a call. The operator can then respond to the correct speaker which is active, replying to the aircraft or ship on the proper frequency.

The circuit diagram for the speaker indicator is shown in Fig. 15-4. A small loudspeaker transformer is connected "backward" to boost the few tenths of a volt up to the 65- to 70-volts dc that are necessary to ionize the neon bulb. The bulb will vary in intensity with the signal being received by that loudspeaker, be it voice, music, or dot-dash Morse code. The parts list for this loudspeaker-energized Telelite is shown in Table 15-2.

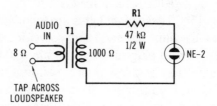

Fig. 15-4. Circuit that will light the Telelite from the audio voltage applied across a loudspeaker.

Table 15-2. Parts List for Speaker-Operated Telelite

Item	Description
NE2	Neon Bulb, $^1/_{25}$-W, Radio Shack 272-1101, or Equivalent
R1	Resistor, 47 kΩ, $^1/_2$-W
T1	Transformer, Audio Output, 1000-Ω Primary, 8-Ω Secondary, Radio Shack 273-1380, or Equivalent
Misc.	Hook-Up Wire, Solder, Tape, Mounting Case

16 Add Music on Hold to Your Telephone

Many of the larger companies and businesses in the country provide music on hold when their customers call in by telephone and there are not sufficient agents to handle their calls. In this chapter, we will cover your adding this feature to your personal phone line.

Background

Many business telephones have had a "hold" feature for a number of years which were provided to all the phones in an office. Buttons on the phone identify each line. When a button was depressed, that light lit with a steady light indicating the line was in use. By pressing a separate button on the telephone instrument the line is placed on hold and the light for that line starts flashing. The flashing light reminded all in the office that that particular line was placed on hold. It would stay on hold, flashing all day long if someone did not pick up the line and answer it. But what about the ones placed on hold? Would they listen to a dead phone for a long while or would they just hang up after a while, perhaps losing a prospective customer or client for that company.

In order to not lose customers and make the wait for service more enjoyable and endurable, many companies went to a feature that provided music to all telephone callers placed on hold. The callers would hear music so long as they were on hold until a person came on the line to handle their transaction. Then the music would be turned off. It was a feature that the

larger businesses could provide but was not readily available to the home-
owner or small business.

Operation

You can have your own music-on-hold for your telephone callers by building
the fairly simple device shown in Fig. 16-1. The MOH (music-on-hold) is built
into a small box or cabinet that is placed alongside your telephone so it can
be easily reached to place it into operation. The MOH has one operating
control—a push button that places the caller on hold. A light emitting diode
(LED) illuminates when the circuit is in operation. An audio input jack
provides the music that will be fed out to your caller. The audio level is
adjusted by varying the control on the radio or tape recorder.

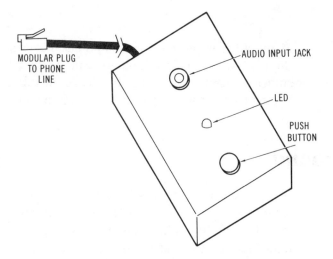

Fig. 16-1. Music-on-hold unit.

Operation of the MOH is as follows. The unit is connected to the telephone
line so that is in parallel with the telephone instrument. When you answer the
phone after it has rung several times, it is in the off-hook condition. However,
you might be in the kitchen and would like to talk for a longer period of time
using the phone in the bedroom or den. You tell your caller that you are going
to place them on hold while you go to another phone to continue the conversation. You would depress the momentary-make push-button switch on the MOH,
hang up the instrument, and go to any other phone in the house and pick up
the phone. The called person will still be on the line. And while you had them
on hold, your MOH played music into their instrument. The music can come

from any AM or FM radio, or from your stereo system if you so like. Once you have picked up any instrument in the house connected to your line, the MOH is removed from the circuit until it is once again required. Its operation is very simple and easy.

Circuit Analysis

Looking at the circuit for the MOH (Fig. 16-2), we see that a total of eight parts makes up the circuit. We will discuss how the circuit works as we walk through each item. You will notice in Fig. 16-2 that the MOH is connected across the telephone line, in parallel with the telephone you have connected across your home or business line. All your local phones are shown on-hook (which means they are not in the circuit at the time, except for the ringer) and the MOH is shown in the activated or on state, which means it is across the phone line.

When your phone is on-hook, 48-volts dc appears across the telephone line at your house or business. This is its normal state, waiting for an

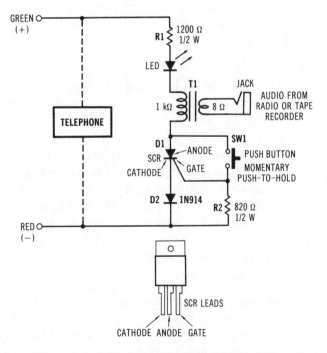

Fig. 16-2. Circuit diagram for Music-on-Hold feature.

incoming call when 90-volts rms will appear on the line to ring your bell circuit. When you answer your phone, the voltage drops to about 6-volts dc. Your phone, which has a dc resistance of about 300-ohms, is placed across the line while you talk. This low resistance tells the telephone central office equipment that your phone is off-hook and a nominal current of 20 to 80 millamperes flows through your circuit.

When you want to activate your MOH, you depress the momentary push-button switch, SW1, and then hang up your phone. This means that your phone is now not across the line. However, your MOH circuit is activated and across the line. The circuit now holds the line so that it does not disconnect from the central office equipment. Music is then fed into your phone line through transformer T1, and your caller hears it as long as he is on hold.

Let's look at the circuit shown in Fig. 16-2 a bit closer. When you press push-button SW1 and hang up your telephone, the line voltage will rise to 48-volts dc because there is no resistance across the line. However, there is a voltage divider formed momentarily by resistor Rl, the LED, transformer T1, and resistor R2. The dc voltage across resistor R1 is applied to the gate of the silicon controlled rectifier (D1). The SCR is triggered into conduction and its resistance drops to a very low value, appearing almost like a short. When you release the push button, the SCR remains conducting so long as its anode is positive and its cathode is connected to ground (negative) through diode D2.

With the SCR acting like a short, diode D2 also acts like a short, connecting resistor R1, LED D1, and transformer T1 across the telephone line. Thus, we have between 1200 and 1500 ohms across the line. This value of resistance is sufficient to place your line on hold. Transformer T1 (an output transformer used backwards) is in series across the line and any audio voltage appearing across the 8-ohm secondary winding will appear across the 1000-ohm primary. This audio voltage is fed back through the telephone line and your caller will hear the music which you are sending him. The 1000-ohm resistance of the primary winding is an ac reactance, usually measured at 1000 hertz. The dc resistance, however, as measured by an ohmmeter, might be as low as a few ohms or tens of ohms.

When you have walked to another part of the house and picked up another telephone, the resistance across the line will drop from the nominal 1200 ohms or so of the MOH to the nominal 300 ohms of your instrument. The line voltage drops to the nominal 6-volts dc with your phone off-hook and there is little voltage across the SCR to keep it on. The SCR shuts off and becomes a high resistance, removing R1, the LED, and the transformer with the audio voltage playing music to the person on hold from the circuit. The

LED goes out because there is no current flow and you can continue your conversation with your caller.

Construction

Table 16-1 shows the parts list for the MOH. These parts are readily available at local radio supply stores. Refer to Fig. 16-1 for layout of the parts on top of the cabinet or case. Push-button SW1 is mounted at the bottom of the case for easy access when placing a caller on hold. The LED above the push button will remain lighted so long as the unit is on hold and playing music. The music, or any commentary you might have tape recorded and wish to play to your callers, is input to the MOH using the audio input jack at the top of the case. A telephone line with modular plug runs out the top edge of the case. Mounting holes for these parts are drilled using a proper size drill. Use sandpaper to smooth off any plastic burrs from the drilling.

Table 16-1. Parts List for Music-On-Hold Circuit

Item	Description
D2	1N914, Radio Shack 276-1122, or Equivalent
Jack	Jack, Two-Conductor, $\frac{1}{8}$-Inch Miniature Phone Jack, Open Circuit, Radio Shack 274-251, or Equivalent
LED	Red or Green LED
SW1	Push-Button, SPST, Momentary Make, Radio Shack 275-1547, or Equivalent
R1	Resistor, 1200-Ω, $\frac{1}{2}$-W, 20%
R2	Resistor, 820-Ω, $\frac{1}{2}$-W, 20%
D1	Silicon Controlled Rectifier, 6-A, 200-V, Radio Shack 276-1067, or Equivalent
T1	Transformer, Audio Output, 1000-Ω Primary, 8-Ω Secondary, Radio Shack 273-1380, or Equivalent
Misc.	Plastic Utility Case, Radio Shack 270-230, or Equivalent; Modular Plug and Cable, Radio Shack 279-310, or Equivalent; Hookup Wire, Solder, Tape

Use a piece of Perf board to mount components that are not mounted on top of the plastic case. Cut out the board to a size that will fit easily inside the case and wedge it in place. Again, to remind yourself as to how much of the circuit you have already connected, mark lightly on the circuit that part which you have already done.

Connection to the Telephone Line

You will probably want to connect the MOH unit into the telephone line at the phone instrument which you answer most often. Use a duplex jack to provide connections for the instrument and the MOH. The MOH should be conveniently reached while holding the phone because you will have to press the push button on the MOH as you hang up the phone. The music is activated the moment you hang up the instrument because this is the time the SCR is triggered into its conduction.

Testing for Operation

Music for the MOH can be obtained from a small radio or tape recorder. Clip two leads across the speaker of the radio and connect them to a miniature plug that is inserted in the jack on top of the MOH. The volume control on the radio or tape recorder is used to control the amount of audio fed to the telephone line through transformer T1. The music is left on as long as you want to provide music service to your callers and can be shut off at night or over weekends.

To set the audio level going out over your line while on hold, set the radio to a comfortable listening level from the loudspeaker. Then call a friend and tell him what you are doing and that you want him to listen and give you an idea of the volume level he receives when you place him on hold. You can do this several times by talking to him and alternately placing him on hold as you adjust the volume level up or down to where it sounds best. Once you find a good level, you can mark a place on the volume control so that you know what level is best. With this final adjustment, you have installed your MOH to your phone which should prove beneficial to your callers when you place them on hold.

17 Design Your Own Audio Amplifier Using Transistors

For many hobbyists, designing transistor amplifiers is a mystery they've never been able to solve. In this chapter, we will briefly review transistor theory and discuss transistor amplifier design. While we cannot cover all of the details of transistor amplifier design in one short chapter, we do believe that enough essential material is included so that a hobbyist with some basic understanding of transistors will be able to design simple, small-signal transistor audio amplifiers after completing this chapter.

What We Will Learn

In this chapter we will cover small-signal (power in the milliwatts) linear amplifiers which employ junction-type transistors. In the discussion on transistor theory, we will limit ourselves to those characteristics and specifications which are needed to understand the operation of a simple transistor amplifier. Atomic structure and the flow of electrons and holes in semiconductor material will not be covered. In some cases, to facilitate understanding, we have reduced complex equations to simple formulas or average values. These simplifications will, on occasion, induce a small error into computations, but the amount of error is usually insignificant compared to the error resulting from using components such as resistors of 5% to 10% tolerance. The same is true for temperature variations, etc.

Transistors

What is a transistor? Simply stated, a transistor is a semiconductor device that controls electric current flow. All transistors have three terminals—a base, an emitter, and a collector. Essentially, the base controls the operation of the transistor. A small change in the base current causes a much greater change in the collector current. Later we will see how this phenomena leads to a voltage gain in the transistor.

Junction-Type Transistors

All junction-type transistors fall into one of two categories—NPN or PNP transistors. Fig. 17-1 shows the schematic symbols for NPN and PNP transistors. For our purpose (without getting into the construction and internal operation of the two categories), we can say that the primary differences between an NPN and a PNP are: (1) the direction of current flow, and (2) the polarity of the emitter, base, and collector voltages.

(A) NPN transistor.   (B) PNP transistor.

Fig. 17-1. Transistor schematic symbols.

Fig. 17-2 illustrates the direction of current flow for NPN and PNP transistors. For the NPN, note that most of the current flows through the emitter to the collector, and a lesser amount of current flows through the emitter to the base. For current to flow in this manner, the collector and base must be positive with respect to the emitter. In the case of PNP transistors, most of the current flows from the collector to the emitter, while a lesser amount of current flows from the base to the emitter. The collector and base must, therefore, be negative with respect to the emitter. Note that in both cases we have said "with respect to the emitter." When analyzing transistor circuits, we should consider the emitter as a reference point because base and collector potentials must be of the proper polarity and within a certain voltage range "with respect to the emitter" before a transistor amplifier can function properly.

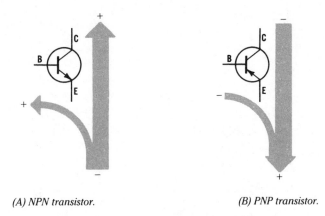

(A) NPN transistor.   (B) PNP transistor.

Fig. 17-2. Current flow in bipolar transistors.

Types of Transistors

Transistors are made from either silicon or germanium semiconductor material. Most modern electronic circuits employ silicon transistors because of their higher current gain characteristics, but some germanium transistors can still be found—usually in circuits built in the '50s and '60s. Silicon Valley is an everyday household word but no one has ever heard of "Germanium Valley!" The type of semiconductor material is important to us because transistors maintain a nearly constant voltage difference between their base and emitter when conducting, and the amount of voltage difference is dependent upon the type of semiconductor material within the transistor. For silicon transistors, the emitter-base voltage difference (E_b – E_e) is about 0.6 to 0.7 volt, and for germaium transistors, the emitter-base voltage difference is usually between 0.15 and 0.35 volt.

Transistor Test Circuits

To demonstrate what this emitter-to-base voltage difference means, we constructed two simple transistor circuits like the one shown in Fig. 17-3. In the first circuit, we used a silicon NPN transistor, and in the second circuit, we used a germanium NPN transistor. We then recorded the emitter voltage (E_e) as the base voltage (E_b) of each amplifier was adjusted from 0 volts to + 0.6 volt. Table 17-1 lists the results of our tests.

Note that the silicon transistor's emitter voltage (Table 17-1) remained at 0 volts when the base voltage was less than + 0.6 volt. For voltages above 0.6 volt, however, the emitter voltage remained at a level roughly 0.6 volt below the base voltage. In effect, the emitter voltage "followed" the base voltage.

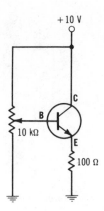

Fig. 17-3. Voltage drop test circuit using NPN transistor.

Table 17-1. Base-Emitter Voltage Difference

Base (E_b)	Silicon Transistor			Germanium Transistor	
	Emitter (E_e)	Difference (E_b − E_e)		Emitter (E_e)	Difference (E_b − E_e)
0.0	0.0			0.0	
0.1	0.0			0.03	0.08
0.2	0.0			0.09	0.11
0.3	0.0			0.18	0.12
0.4	0.0			0.27	0.13
0.5	0.0			0.36	0.14
0.6	0.04	0.56		0.46	0.14
0.8	0.22	0.58		0.65	0.15
1.0	0.41	0.59		0.85	0.15
2.0	1.40	0.60		1.84	0.16
3.0	2.40	0.60		2.84	0.16
4.0	3.39	0.61		3.83	0.17
5.0	4.39	0.61		4.83	0.17
6.0	5.38	0.62		5.82	0.18

In the case of the germanium transistor, the emitter voltage was also 0 volt when the base voltage was 0 volt. As the base voltage increased from +0.2 to +0.6 volt, the emitter voltage followed the base voltage, but in this case the voltage difference was roughly 0.15 volt.

So What's the Difference?

Knowing the approximate emitter-base voltage difference ($E_b - E_e$) for silicon and germanium transistors is very important when designing or analyzing transistor amplifiers. For our purpose, we will assume that the emitter-base voltage drop will be about 0.65 volt for silicon transistors and about 0.2 volt for germanium transistors. Based upon personal experience, we prefer to use 0.2 volt (instead of 0.15 volt) in our calculations for germanium transistors because it arises more often.

Beta

Transistors are current amplifying devices. There are thousands of different types, and each type has its own unique set of specifications, such as, maximum operating voltage (Max V_{ce}), maximum power dissipation, maximum collector current (Max I_c), etc. One other specification is of special interest, and that is the current amplification factor, or *Beta* (often shown as h_{FE} on a transistor specification sheet). Beta is the ratio of collector current to base current, or Beta = I_c/I_b. For newer types of transistors, Beta is often 4 times greater than the base current when the transistor collector current is at a level specified by the manufacturer. For example, the Beta of a 2N2222 transistor (a commonly available type) is typically about 200 when the collector current is 150 mA. In reality, a transistor is rarely operated at the optimum collector current level that provides maximum Beta. So, for our purposes, we generally use a Beta value that is only 50% of the value identified by the manufacturer. At first glance, you may believe that such a generalization for Beta could induce significant error into our calculations, but later you will see that the overall affect is minimal.

A Basic Transistor Amplifier

To illustrate how a transistor amplifier works, we constructed a simple test amplifier as shown in Fig. 17-4. A 2N2222 transistor was selected for the test because it is a commonly used and inexpensive silicon device with relatively high current gain characteristics. Many other types of transistors could have been used in place of the 2N2222.

Test Setup

In this test, voltage and current measurements will be recorded as potentiometer R1 is adjusted from 0 volts upwards until the transistor reaches saturation (maximum collector current). Table 17-2 lists the base voltage (E_b), base current (I_b), emitter voltage (E_e), emitter current (I_e), collector voltage (E_c), and collector current (I_c).

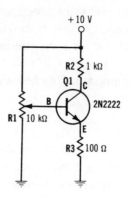

Fig. 17-4. Test amplifier configuration using NPN transistor and positive power supply.

Table 17-2. Transistor Amplifier Voltage and Current Measurements

Base		Emitter		Collector	
E_b (V)	I_b (μA)	E_e (V)	I_e (mA)	E_c (V)	I_c (mA)
0.0	0	0.0	0.0	10.0	0.0
0.1	0	0.0	0.0	10.0	0.0
0.2	0	0.0	0.0	10.0	0.0
0.3	0	0.0	0.0	10.0	0.0
0.4	0	0.0	0.0	10.0	0.0
0.5	1	0.005	0.05	9.95	0.05
0.6	5	0.04	0.41	9.60	0.40
0.7	15	0.12	1.3	8.75	1.25
0.8	17	0.21	2.1	7.90	2.10
0.9	22	0.30	3.0	7.05	2.95
1.0	27	0.38	3.8	6.20	3.80
1.1	32	0.47	4.7	5.35	4.65
1.2	37	0.55	5.5	4.50	5.50
1.3	42	0.64	6.4	3.65	6.35
1.4	47	0.73	7.3	2.80	7.20
1.5	52	0.81	8.1	1.95	8.05
1.6	69	0.90	9.0	1.11	8.90
1.7	760	0.97	9.7	1.05	8.95

Before reviewing the data in Table 17-2, carefully study the schematic in Fig. 17-4. Transistor Q1 is an NPN, and the power supply provides +10 volts. We, therefore, can expect Q1's collector and base voltages to be positive with respect to the emitter. Potentiometer R1 varies the bias voltage applied to the base of Q1. As the base voltage is increased, we should expect the emitter voltage to follow the base, but at a level about 0.65 volt below the base voltage. As the emitter voltage increases, so will the current passing through emitter resistor R3. Q1's emitter current and collector current will also increase because of the increase in the current through R3. As collector current increases, the voltage drop across collector resistor R2 will also increase. Thus, an increase in the base voltage will give us a decrease in collector voltage.

Beta for a 2N2222 is typically about 200 when collector current is 150 mA. For reasons stated earlier, we assume the worst case and estimate Beta at 50% of its rated value, or about 100. This means that the amount of current passing through the collector will be roughly 100 times greater than the base current.

Test Results

Let's refer again to the data in Table 17-2. We will now procede through our discussion of how the test amplifier of Fig. 17-4 works. You will see that by using what we have learned thus far, coupled with Ohm's Law, we are able to calculate the voltage and current at the test amplifier's emitter, base, and collector, as well as its voltage gain. We'll compare our calculations to the actual measurements listed in Table 17-2. In the table, notice that potentiometer R1 was adjusted from 0 to +1.7-volts in 0.1-volt steps (measured at the base of Q1). The transistor was in a "cut-off" state when the base voltage was less than +0.5 volt. Cut-off means that no current is flowing through the transistor's base, emitter, or collector. Since no current is flowing, the collector voltage equals V_{cc} (+10 volts) and the emitter voltage is zero.

The transistor conducts slightly when the base voltage is adjusted to between +0.5 and +0.6 volt. You can see that base, emitter, and collector current are flowing, and that the collector voltage decreases slightly. When the base voltage is +1.7 volts, the emitter and collector voltages are almost equal. Maximum conduction occurs and the transistor is in a "saturation" state. Adjusting the base voltage above +1.7 volts causes excessive base current and the transistor can be damaged.

The voltage range from +0.7 and +1.5 volts can be considered the operating range of the amplifier because this range of input voltages provides us with a linear amplified voltage at the collector output. By this we mean that for each 0.1-volt increase in base voltage, we had a corresponding 0.85-volt decrease at the collector. The voltage gain of this amplifier is calculated to be 8.5 (0.85 volt/0.1 volt). Later you will see that the voltage gain of the amplifier can be determined by the values of R2 and R3, and not by the Beta of the transistor.

Test Analysis

Let us now perform a step-by-step analysis of the test amplifier in Fig. 17-4. In this analysis, we will compute the current and voltage for the test amplifier base, emitter, and collector. We will then compare our computations to the actual values listed in Table 17-2. Our analysis will be at one specific base voltage (+1.0 volts), but would be applicable at any other base voltage within the operating range of +0.7 and +1.5-volts.

1. A base voltage of +1.0 volt has been selected for this analysis. The emitter voltage must be about 0.65-volt below the base voltage, or

$$E_e = E_b - 0.65 = 1.0 - 0.65 = 0.35 \text{ volt.}$$

    Comparing our calculations for E_e with the measurement listed in Table 17-2 (0.38 volt), we see that the calculated value is off by only 0.03 volt, an error of about 9%.

2. From Step 1, we know that the emitter voltage must be about +0.35 volt. The emitter current is equal to the current that passes through emitter resistor R3. If the voltage drop across R3 is 0.35 volt, then the current through R3 (and the emitter of Q1) must be

$$I_e = E/R = 0.35/100 = 3.5 \text{ mA.}$$

    Again, by comparing our calculation for I_e with the measured value in Table 17-2, we find that our calculation was off by only 0.3 mA, an error of about 9%.

3. As stated earlier, we expect Beta to be at least 100 for the 2N2222 used in our amplifier. This means that the collector current should be about 100 times greater than the base current, or about 99% of the emitter current will pass through the collector and 1% should pass through the base. Based on that assumption, collector current and base current should be about

$$I_c = 99\% \times I_e = 0.99 \times 3.5 = 3.465 \text{ mA}$$
and,
$$I_b = 1\% \times I_e = 0.01 \times 3.5 = 0.035 = 35 \text{ } \mu A.$$

    Comparing these calcualtions with the values listed in Table 17-2 for I_c (3.8 mA) and I_b (27 μA), we see that the calculations are off by 0.035 mA and 8 μA, respectively.

4. The amount of current flowing through collector resistor R2 is the same as the collector current. Consequently, the voltage drop across R2 must be

$$E = IR = 3.465 \times 1000 = 3.465 \text{ volts.}$$

The collector voltage is simply the voltage dropped across R2 subtracted from the power supply voltage (V_{cc}), or

$$E_c = V_{cc} - E = 10 - 3.465 = 6.535 \text{ volts.}$$

Comparing this calculation for E_c with the value listed in Table 17-2, we see that the calculation is off by 0.335 volt, an error of about 5%. We have calculated the base current, emitter voltage, emitter current, collector voltage, and collector current for the test amplifier in Fig. 17-4 when the base voltage was + 1.0 volt. Obviously, this procedure allows us to calculate these values for any other base voltages within the operating range of + 0.7 and + 1.5 volts.

Accuracies Expected

How accurate should we expect our calculations to be? For the test amplifier in Fig. 17-4 and Table 17-2, all of the calculations (with the exception of the base current) were within 9%. Several factors induce error into these calculations, such as:

1. Resistors will induce some error because the tolerances of the resistors used in amplifiers are typically 5% or 10%.

2. Significant increases or decreases in the amount of current flowing through a transistor can cause the temperature of the transistor to increase or decrease. These changes in temperature, in turn, cause the base-emitter voltage difference, Beta, and other transistor characteristics, to change.

3. The accuracy of the meter used for measurements may induce error. The meter used for the above test has a full-scale accuracy of ± 3%.

4. In some cases, we used estimated values instead of exact values for the calculations. These estimations, quite obviously, will induce some error.

Taking all of these sources of error into account, we could, on occasion, find that our calculations may be off by 15%, or more. In practice, however, you'll probably find that your calculated values will often be within 10%.

Voltage Gain Calculation

Depending on circuit complexity, calculating the voltage gain of a transistor amplifier can be relatively simple or very difficult. In our discussion on voltage gain, we will limit ourselves to those amplifiers that have noninductive base, emitter, and collector circuits. Adding reactive components (other than bypass and coupling capacitors) greatly compounds gain calculations. Accordingly, we will limit our discussion to the simple amplifier.

The formula for determining voltage gain of a small-signal transistor amplifier is

$$\text{Gain} = R_c/(R_e + R_e')$$

where,
  R_c is the total ac collector resistance,
  R_e is the total external emitter ac resistance,
  R_e' is the internal emitter resistance of the transistor.

All transistors have an internal emitter resistance. The amount of this resistance depends upon the operating temperature of the transistor and the amount of emitter current. For most small-signal transistor amplifiers, R_e' is usually between one ohm and several hundred ohms. The internal emitter resistance of a transistor operating at room temperature can be computed using the formula

$$R_e' = 0.025/I_e.$$

Referring to the test amplifier of Fig. 17-4 and our analysis in the preceding section, we estimate that I_c will be about 3.5 mA when the base bias voltage is adjusted to +1.0 volt. Using that value, R_e' will be about

$$R_e' = 0.025/0.0035 = 7.1 \text{ ohms.}$$

Applying this value to the voltage gain formula, we have

$$\begin{aligned} \text{Gain} &= R_c/(R_e + R_e') \\ &= 1000/(100 + 7.1) \\ &= 9.3. \end{aligned}$$

The actual gain for the test amplifier in Fig. 17-4 was 8.5. Our estimate of 9.3 is off by just 9%.

More Circuit Analysis

Notice that the amplifier in Fig. 17-5 uses a 2N1305 (PNP) transistor and that the supply voltage is reversed, providing -15 volts. The 2N1305 is a germanium transistor with a Beta of about 40. Resistors R1 and R2 form a voltage divider network that establishes the base voltage for the amplifier. Capacitor C1 is an input coupling capacitor that couples the input signal to the base of the transistor. R3 is the collector resistor, and C2 couples the output signal to the 10-kΩ output load. Resistors R4 and R5 are the external emitter resistors. R4 and R5 together affect the dc (static) operation of the amplifier, but only resistor R4 will affect the ac (dynamic) operation of the amplifier because capacitor C3 acts as a short (about zero ohms) for ac signals.

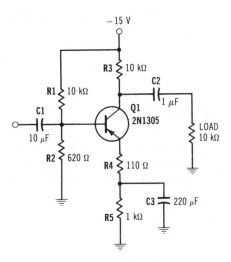

Fig. 17-5. Transistor audio amplifier using PNP transistor and negative power supply.

Static Analysis

First, let's analyze the amplifier under static conditions when no ac signal is applied. A PNP transistor is being used, and the power supply provides -15 volts, so all voltages must be negative. The 2N1305 is a germanium transistor, so the base-emitter voltage drop will be about 0.2 volt. Capacitors have no affect on dc circuits so C1, C2, C3, and the load resistor can be ignored.

Let's step through the static analysis and determine E_b, E_e, and E_c.

1. Base voltage: Disregard the current that passes through R1 and the base of the transistor. The voltage at the junction of R1 and R2 can then be estimated as follows:

$$I = E/(R1 + R2)$$
$$= -15/(10,000 + 620)$$
$$= -1.41 \text{ mA}$$

and,

$$E_b = I \times R2$$
$$= -0.00141 \times 620$$
$$= -0.87 \text{ volts}$$

In reality, the base voltage would be slightly less than -0.87 volt because the base current must also flow through R1. For most amplifiers, however, this procedure will provide a close estimate.

2. Emitter voltage: Since a germanium transistor is used, the emitter voltage should be

$$E_e = E_b - 0.2 \text{ volt}$$
$$= -0.87 - (0.2)$$
$$= -0.67 \text{ volt}$$

3. Collector current: Since we assume that Beta will be approximately 50% of its rated value, Beta for this amplifier is assumed to be 20. This means that about 5% $(1/20 = 5\%)$ of the emitter current will flow through the base and 95% will flow through the collector. First we calculate the emitter current as

$$I_e = E/(R4 + R5)$$
$$= -0.67/(110 + 1000)$$
$$= -0.6 \text{ mA}$$

and then the collector current,

$$I_c = 95\% \times I_e$$
$$= 0.95 \times (0.6)$$
$$= 0.57 \text{ mA.}$$

4. Collector voltage: The collector voltage is equal to the power supply voltage (V_{cc}) minus the voltage drop across collector resistor R3, or

$$E_c = V_{cc} - (I_c \times R3)$$
$$= -15 - (-0.00057 \times 10,000)$$
$$= -15 - (5.7)$$
$$= -9.3 \text{ volts.}$$

Dynamic Analysis

Now that we've completed a static analysis of the amplifier and calculated the values of E_b, E_e, and E_c, let's perform a dynamic analysis of the amplifier in Fig. 17-5. Assume that a 1000-Hz, 0.1-volt, peak-to-peak (p-p) signal is applied to the input coupling capacitor C1. For the purposes of our analysis, assume that all capacitors act as ac shorts. The signal level at the base of Q1 must, therefore, also be 0.1 volt p-p. Since capacitor C3 is connected to ground, R5 is effectively removed from the circuit (it is shorted to ground). Only R4 and the internal resistance (R_e') of the transistor will affect the gain of the amplifier.

In the collector circuit, capacitor C2 acts as an ac short. This effectively places the load resistor (10 kΩ) in parallel with R3 (10 kΩ). Recall that the gain formula is

$$\text{Gain} = R_c/(R_e + R_e').$$

Since R3 and the load resistors are in parallel, then

$$
\begin{aligned}
R_c &= (R3 \times \text{load})/(R3 + \text{load}) \\
&= (10,000 \times 10,000)/(10,000 + 10,000) \\
&= 5000 \text{ ohms.}
\end{aligned}
$$

R_e is equal to R4 (110 ohms) because capacitor C3 shorts the junction of R4 and R5 to ground. R_e' is computed using the formula mentioned earlier. That is,

$$
\begin{aligned}
R_e' &= 0.025/I_e \\
&= 0.025/0.0006 \\
&= 42 \text{ ohms.}
\end{aligned}
$$

The voltage gain of the amplifier in Fig. 17-5 is then shown to be

$$
\begin{aligned}
\text{Gain} &= R_c/(R_e + R_e') \\
&= 5000/(110 + 42) \\
&= 33.
\end{aligned}
$$

We built an amplifier like the one shown in Fig. 17-5 and measured E_b, E_e, E_c, and its voltage gain. The measured results are shown in Table 17-3.

Input and Output Impedance

Before designing a transistor amplifier, we must look at two other important characteristics of amplifiers—input and output impedance.

Table 17-3. Comparison of Calculated Value to Actual Values
for the Amplifier in Fig. 17-5

Measurement	Calculated Value	Actual Value
E_b	− 0.87 V	− 0.85 V
E_e	− 0.67 V	− 0.66 V
E_c	− 9.3 V	− 8.4 V
Gain	33	34

Background

Impedance is defined as the ac resistance of an electronic device or circuit. If, for example, a microphone with an output impedance of 600 ohms is to be connected to the input of a transistor amplifier, we would want the input impedance of the amplifier to be at least 600 ohms so that most of the signal would be passed to the base of the transistor. If the input impedance of the amplifier were much lower, say 100 or 200 ohms, then the output of the microphone would be severely loaded, reducing the amplitude of the input to the base of the transistor.

Just as input impedance is very important, output impedance is also critical to an amplifier's operation. For example, if a transistor amplifier with an output impedance of 1000 ohms were connected to an 8-ohm speaker, the impedance mismatch would cause a significant reduction in output amplitude. This mismatch might even damage the transistor amplifier.

When designing amplifiers, we try to insure that the input impedance of a device (or amplifier stage) closely matches the output impedance of the previous device or stage. As will be seen in the next section, it is sometimes acceptable to connect a device or circuit with a high input impedance to a device or circuit with a low output impedance, but the reverse should be avoided.

Output Impedance

To determine the output impedance of a small-signal transistor amplifier, simply look at the collector resistor. The output impedance of the amplifier is approximately equal to the value of the collector resistor. Referring to Fig. 17-5, the output impedance of that amplifier is approximately 10 kΩ (note that the load resistance selected for this example was, in fact, 10 kΩ).

Input Impedance

Determining the input impedance of a transistor amplifier is not quite as easy. Referring once again to Fig. 17-5, we can see that bias resistors R1 and R2 and the base of the transistor are connected to the input capacitor C1 and will affect the input impedance. Effectively, all three are in parallel, so that the total input impedance will be

$$1/Z_{in} = 1/R1 + 1/R2 + 1/Z_b,$$

where,

Z_{in} is the input impedance of the circuit,
Z_b is the base input impedance.

The base imput impedance can be estimated by the formula:

$$Z_b = Beta \times (R_e + R_e').$$

Thus, for the amplifier in Fig. 17-5 Z_b is calculated to be:

$$Z_b = 20 \times (110 + 42)$$
$$= 3040 \text{ ohms,}$$

and Z_{in} can be determined by

$$1/Z_{in} = 1/R1 + 1/R2 + 1/Z_b$$
$$= 1/10,000 + 1/620 + 1/3040$$
$$1/Z_{in} = 0.00204,$$

or,

$$Z_{in} = 1/0.00204$$
$$= 490 \text{ ohms.}$$

Transistor Amplifier Design

We will now go through a step-by-step procedure which will enable you to design small-signal transistor audio amplifiers. The step-by-step procedure outlined in this chapter is different than the one used used by an experienced design engineer. Our intent is to simplify the design process for the hobbyist, not to teach all that is to be known about transistor amplifier design. The following procedure places upper and lower limits on most specifications and also uses several simplified equations. By stating limits and using these equations, we will guide you through what is normally a very complex design process.

The Steps

The steps in designing the small-signal transistor audio amplifier are outlined in Table 17-4. In Step 1 of the procedure, the specifications of the
amplifier are determined. As stated in the previous paragraph, upper and
lower limits are identified to help keep you within the required limits.
Amplifiers can be designed that exceed these limits, but their design is
much more complex. In Steps 2, 3, and 4, the collector, emitter, and base
circuits are designed. Step 5 identifies several points to consider when
constructing amplifiers that have been designed using this procedure.

Table 17-4. Transistor Amplifier Design Procedure

Step	Description	Values
1	Determine Amplifier Specification	
a	Lowest frequency to be amplified	LF = _____ Hz
	Must be less than 10 Hz	
b	Peak-to-Peak Output Voltage (E_{out})	E_{out} = _____ V (p-p)
	E_{out} must not exceed 15 V (p-p)	
c	Power Supply Voltage (V_{cc})	V_{cc} = _____ V
	V_{cc} must be between 6 and 30V	
	Minimum V_{cc} = 2 × E_{out}	
	If minimum V_{cc} is less than 6V, then let minimum V_{cc} equal 6V	
	Select power supply voltage between minimum V_{cc} and 30V	
	Select polarity of V_{cc}	+/− = _____
d	Select Transistor Type	Type = _____
	If V_{cc} is positive, select NPN	
	If V_{cc} is negative, select PNP	
	Select transistor with a Max V_{ce} that is greater than V_{cc}	
	Consider using 2N2222 (NPN with h_{FE} of 200) or 2N3906 (PNP with h_{FE} of 50)	
	Let Beta equal 50% of selected transistor h_{FE}	Beta = _____
e	Output Load (equal to input impedance of next circuit or device)	Load = _____ ohms
	Load must not be less than 1000 ohms	
f	Output Impedance (Z_{out})	Z_{out} = _____ ohms
	Z_{out} must be between 1000 and (330 × V_{cc}) ohms	
	If load is less than (330 × V_{cc}) ohms, then let Z_{out} equal load	
	If load is greater than (330 × V_{cc}) ohms, then let Z_{out} equal (330 × V_{cc}) ohms	
g	Desired Voltage Gain	Gain = _____

Table 17-4. cont.

Step	Description	Values
	Desired Gain must be between 2 and Max Gain, where $$\text{Max Gain} = \frac{(5 \times V_{cc})}{[(Z_{out}/\text{Load}) + 1]}$$	
h	Input Impedance (Z_{in}) Z_{in} must be less than Max Z_{in}, where $$\text{Max } Z_{in} = \frac{(\text{Beta} \times Z_{out})}{[(2 \times \text{Gain}) + 10]}$$ If desired Z_{in} is greater than Max Z_{in}, then reduce desired gain or select a transistor with a higher h_{FE}, then return to step 1a and restart the design procedure	Z_{in} = _____ ohms
2	Collector Circuit Design	
a	Collector Resistor (R3) Select standard resistor that is close to Z_{out} (Step 1f)	R3 = _____ ohms
b	Output Coupling Capacitor (C2) Value of C2 is in microfarads LF is the lowest frequency to be amplified (Step 1a) Select standard value capacitor close to the computed value for C2, where C2 = 1,590,000/(LF × Load)	C2 = _____ μF
c	Collector Voltage (E_c) $E_c = 0.66 \times V_{cc}$	E_c = _____ V
d	Collector Current (I_c) $I_c = (V_{cc} - E_c) / R3$	I_c = _____ mA
3	Emitter Circuit Design	
a	Emitter Current (I_e) $I_e = I_c + (I_c/\text{Beta})$	I_e = _____ mA
b	Internal Emitter Resistance (R_e') $R_e' = 0.025/I_e$	R_e' = _____ ohms
c	Emitter Resistor (R4) $R_c = (R3 \times \text{Load})/(R3 + \text{Load})$ $R4 = (R_c/\text{Gain}) - R_e'$ If value of R4 is less than 0 ohm, then design process must be restarted at Step 1a using a lower desired voltage gain or a higher desired output impedance If the computed value for R4 is greater than 0 ohm, select a standard value resistor close to the computed value for R4	R4 = _____ ohms
d	Skip Steps 3e and 3f if ($I_e \times R4$) is equal to, or greater than 10% of V_{cc}	
e	Emitter Resistor R5 $R5 = [(0.1 \times V_{cc}) - (I_e \times R4)] / I_e$ Select a standard value resistor close to the computed value for R5	R5 = _____ ohms

(cont. on next page)

Table 17-4. cont.

Step	Description	Values
f	Bypass Capacitor (C3) Value of C3 is in microfarads LF is the lowest frequency to be amplified (Step 1a) Select a standard value capacitor close to the value computed for C3, where $C3 = 1,590,000/[LF \times (R4 + R_e')]$	$C3 =$ _____ μF
g	Emitter Voltage (E_e) $E_e = I_e \times (R4 + R5)$	$E_e =$ _____ V
4	Base Circuit Design	
a	Base Voltage (E_b) If silicon transistors are used, then let $E_b = E_e + 0.65$-V If germanium transistors are used, then let $E_b = E_e + 0.2$-V	$E_b =$ _____ V
b	Base Current (I_b) $I_b = I_c/Beta$	$I_b =$ _____ mA
c	Bias Resistor (R2) $R2 = Z_{in} \times [1.5 + (0.3 \times Gain)]$ Select a standard value resistor close to the computed value for R2	$R2 =$ _____ ohms
d	Bias Resistor (R1) $R1 = (V_{cc} - E_b)/[I_b + (E_b/R2)]$ Select a standard value resistor close to the computed value for R1	$R1 =$ _____ ohms
e	Input Coupling Capacitor (C1) Value of C1 is in microfarads LF is the lowest frequency to be amplified (Step 1a) Select a standard value capacitor close to the computed value for C1, where $C1 = 1,590,000/(LF \times Z_{in})$	$C1 =$ _____ μF
5	Amplifier Construction If the polarity of V_{cc} (Step 1c) is negative, then V_{cc}, E_c, E_e, and E_b will be negative when the amplifier is constructed. When constructing the amplifier, the value of R1 may have to be increased or decreased slightly to make E_c equal to $0.66 \times V_{cc}$. Computed values for capacitors are very conservative. In many cases, a capacitor several times smaller than the computed value will work well with minimal effect. The working voltage of all capacitors should be greater than V_{cc}. Use resistors that have a wattage rating that is at least two times the power dissipated (by the resistor). The power dissipated can be determined by the formula $P = I \times E$	

Examples of the Design Process

To demonstrate the use of the design process, assume that a transistor amplifier is needed that will amplify an audio signal. The input device has an impedance of 1000 ohms and the circuit (or device) to be driven by the amplifier has an impedance of 10 kΩ. The input voltage will be 0.1 volt p-p, and the output voltage must be 1 volt p-p (a voltage gain of 10). Assume that a -12-volt power supply is available, and the minimum frequency to be amplified is 20 Hz.

One-Stage Audio Amplifier

Table 17-5 lists the results of the computations. A 2N3906 transistor was selected because a PNP transistor was required. A transistor with a higher h_{FE} would have been required if the desired input impedance (Z_{in} in Step 1h) was greater than the maximum allowable input impedance (Max Z_{in}). In general, a transistor with a high h_{FE} (100 or more) should be used if the desired gain is high and the desired input impedance is close to (or greater than) the output impedance.

Table 17-5. Computations for a Transistor Amplifier

Step		Results	Step		Results
1	a	LF = 20 Hz	3	a	I_e = 1.09 mA
	b	E_{out} = 1 V (p-p)		b	$R_e{}'$ = 23 ohms
	c	V_{cc} = 12 V		c	R4 = 240 ohms
		Polarity = Neg			(R_c = 2805 ohms)
		Min V_{cc} = 6 V			(Calculated value = 257 ohms)
	d	Type = 2N3906		d	Go through Steps 3e and 3f
		Beta = 25		e	R5 = 820 ohms
	e	Load = 10,000 ohms			(Calculated value = 820 ohms)
	f	Z_{out} = 3900 ohms		f	C3 = 470 μF
		Max Z_{out} = 3960 ohms			(Calculated value = 302 μF)
	g	Gain = 10		g	E_e = 1.15 V
		Max gain = 43	4	a	E_b = 1.81 V
	h	Z_{in} = 1000 ohms		b	I_b = .042 mA
		Max Z_{in} = 3250 ohms		c	R2 = 4300 ohms
2	a	R3 = 3900 ohms			(Calculated value = 4500 ohms)
	b	C2 = 10 μF		d	R1 = 22,000 ohms
		(Calculated value = 7.9 μF)			(Calculated value = 22,000 ohms)
	c	E_c = 7.92 V		e	C1 = 100 μF
	d	I_c = 1.05 mA			(Calculated value = 80 μF)

Note that V_{cc}, E_c, E_e, and E_b are listed as positive voltages in Table 17-5. To simplify the design process, we consider all voltages to be positive when making computations. But, when constructing an amplifier that has a negative power supply voltage (V_{cc}), you should expect E_c, E_e, and E_b to also be negative.

Fig. 17-6 shows the schematic for the amplifier. Note that Step 3d of the design procedure states that Steps 3e and 3f should be skipped if $I_e \times R4$ is greater than 10% of V_{cc}. If Steps 3e and 3f are to be skipped in the design process, then the amplifier will not have resistor R5 and capacitor C3. The bottom of resistor R4 would then be connected to ground.

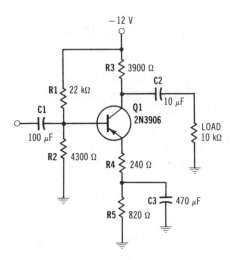

Fig. 17-6. One-stage audio amplifier using PNP transistor and negative power supply.

Two-Stage Audio Amplifier

As a second example, assume that a transistor amplifier is needed to amplify an audio signal from 10 mV (p-p) up to 1 volt (p-p) (a voltage gain of 100). The impedance of the input device is 1000 ohms and the circuit to be driven has an input impedance of 1000 ohms. A 9-volt battery will be used for power. The lowest frequency to be amplified is 20 Hz.

If you started going through the design procedure, you would get as far as Step 1g before stopping. At that point you would realize that the desired gain (100) exceeded the maximum allowable gain (22). Now what? The answer is to design a two-stage amplifier. If each stage of a two-stage amplifier provides a gain of 10, then the total gain of the amplifier is 10 × 10 or 100.

Table 17-6 lists the computations for the second and the first stage of the amplifier. Note that the second stage is designed first. Also note that the input impedance selected for the second stage is the maximum allowable input impedance (Max Z_{in}). The input impedance (Z_{in}) of the second stage is then used as the load for the first stage. Capacitor C2 is not required in the first stage because capacitor C1 of the second stage is used as a coupling capacitor.

Table 17-6. Computations for a Two-Stage Transistor Amplifier

Step	Results	Step	Results
Stage 2			
1 a	LF = 20 Hz	3 a	I_e = 3.09 mA
b	E_{out} = 1 V (p-p)	b	R_e' = 8.1 ohms
c	V_{cc} = 9 V	c	R4 = 43 ohms
	Polarity = Pos		(R_c = 500 ohms)
	(Min V_{cc} = 6 V		(Calculated value = 42 ohms)
d	Type = 2N2222	d	Go through Steps 3e and 3f
	Beta = 100	e	R5 = 240 ohms
e	Load = 1000 ohms		(Calculated value = 248 ohms)
f	Z_{out} = 1000 ohms	f	C3 = 2000 μF
	Max Z_{out} = 2970 ohms		(Calculated value = 1555 μF)
g	Gain = 10	g	E_e = 0.87 V
	Max gain = 22	4 a	E_b = 1.52 V
h	Z_{in} = 3333 ohms	b	I_b = 0.031 mA
	Max Z_{in} = 3333 ohms	c	R2 = 15,000 ohms
2 a	R3 = 1000 ohms		(Calculated value = 14,999 ohms)
b	C2 = 100 μF	d	R1 = 56,000 ohms
	(Calculated value = 79.5 μF)		(Calculated value = 56,524 ohms)
c	E_c = 5.94 V	e	C1 = 50 μF
d	I_c = 3.06 mA		(Calculated value = 24 μF)
Stage 1			
1 a	LF = 20 Hz	3 a	I_e = 1.03 mA
b	E_{out} = 0.1 V (p-p)	b	R_e' = 24 ohms
c	V_{cc} = 9 V	c	R4 = 130 ohms
	Polarity = Pos		(R_c = 1587 ohms)
	(Min V_{cc} = 6 V		(Calculated value = 135 ohms)
d	Type = 2N2222	d	Go through Steps 3e and 3f
	Beta = 100	e	R5 = 750 ohms
e	Load = 3333 ohms		(Calculated value = 744 ohms)
f	Z_{out} = 2970 ohms	f	C3 = 470 μF
	Max Z_{out} = 2970 ohms		(Calculated value = 516 μF)
g	Gain = 10	g	E_e = 0.91 V
	Max gain = 23	4 a	E_b = 1.56 V
h	Z_{in} = 1000 ohms	b	I_b = 0.010 mA
	Max Z_{in} = 9900 ohms	c	R2 = 4300 ohms
2 a	R3 = 3000 ohms		(Calculated value = 4400 ohms)
b	C2 is same capacitor as C1 in Stage 2	d	R1 = 20,000 ohms
			(Calculated value = 20,578 ohms)
c	E_c = 5.94V	e	C1 = 100 μF
d	I_c = 11.02 mA		(Calculated value = 79.5 μF)

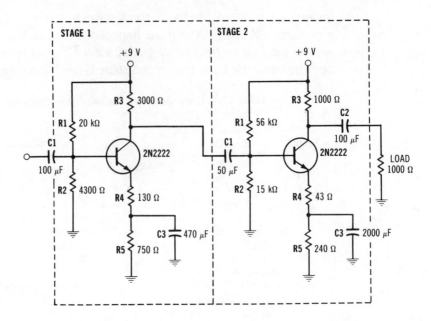

Fig. 17-7. Two-stage audio amplifier using NPN transistors and positive power supply.